What Water Is Worth

DOI: 10.1057/9781137062499

Also by Zachary A. Smith

THE ENVIRONMENTAL POLICY PARADOX (*fifth edition, 2009*)

GLOBALIZATION: A Reference Handbook (*with Justin Ervin, 2008*)

RENEWABLE AND ALTERNATIVE ENERGY RESOURCES (*with Katrina Taylor, 2008*)

ENVIRONMENTAL POLITICS AND POLICY IN THE WEST (*edited with John Freemuth, 2007*)

PROTECTING OUR ENVIRONMENT: Lessons from the European Union (*with Janet R. Hunter, 2005*)

THE ENVIRONMENTAL POLICY PARADOX (*fourth edition, 2004*)

OCEAN POLITICS AND POLICY (*with Peter Jacques, 2003*)

FRESHWATER ISSUES (*with Grenetta Thomassey Fink, 2003*)

THE NATIONAL ENVIRONMENTAL POLICY ACT: Judicial Misconstruction, Legislative Indifference & Executive Neglect (*with Matt Lindstrom, 2001*)

HAWAI'I POLITICS AND GOVERNMENT: An American State and a Pacific Society (*with Richard Pratt, 2000*)

THE ENVIRONMENTAL POLICY PARADOX (*third edition, 2000*)

GROUNDWATER MANAGEMENT IN THE WEST (*with Jeff Ashley, 1999*)

PUBLIC POLICY AND POLITICS IN ARIZONA (*second edition, editor, 1996*)

THE ENVIRONMENTAL POLICY PARADOX (*second edition, 1995*)

ENVIRONMENTAL POLITICS AND POLICY IN THE WEST (*editor, 1993*)

PUBLIC POLICY AND POLITICS IN ARIZONA (*editor, 1993*)

THE ENVIRONMENTAL POLICY PARADOX (*1992*)

POLITICS AND PUBLIC POLICY IN HAWAI'I (*edited with Richard Pratt, 1992*)

ALOHA COUNTY: The Hawai'i Local Government Simulation (*1992*)

WATER AND THE FUTURE OF THE SOUTHWEST (*editor, 1989*)

GROUNDWATER IN THE WEST (*1989*)

GROUNDWATER POLICY IN THE SOUTHWEST (*1985*)

INTEREST GROUP INTERACTION AND GROUNDWATER POLICY FORMATION IN THE SOUTHWEST (*1985*)

DOI: 10.1057/9781137062499

palgrave▸pivot

What Water Is Worth: Overlooked Non-Economic Value in Water Resources

Kira Artemis Russo

and

Zachary A. Smith

palgrave macmillan

DOI: 10.1057/9781137062499

First published 2013 by
PALGRAVE MACMILLAN

Palgrave Macmillan in the UK is an imprint of Macmillan Publishers Limited, registered in England, company number 785998, of Houndmills, Basingstoke, Hampshire RG21 6XS.

Palgrave Macmillan in the US is a division of St Martin's Press LLC, 175 Fifth Avenue, New York, NY 10010.

Palgrave Macmillan is the global academic imprint of the above companies and has companies and representatives throughout the world.

Palgrave® and Macmillan® are registered trademarks in the United States, the United Kingdom, Europe and other countries

ISBN: 978-1-137-16526-8 EPUB
ISBN: 978-1-137-06249-9 PDF
ISBN: 978-0-230-34076-3 Hardback

This book is printed on paper suitable for recycling and made from fully managed and sustained forest sources. Logging, pulping and manufacturing processes are expected to conform to the environmental regulations of the country of origin.

A catalogue record for this book is available from the British Library.

A catalog record for this book is available from the Library of Congress.

www.palgrave.com/pivot

DOI: 10.1057/9781137062499

I dedicate this work to my boys, Alex and Andy, in the hope that their generation will not suffer at the hands of the past, to Mother and Dad for their lifelong love and support, and to my partner, Winter, as a reminder that love always prevails.

With this, I offer myself to Thee ...

Kira Artemis Russo

For Alden and Genevieve, the light of my life

Zachary A. Smith

DOI: 10.1057/9781137062499

With the arrival of the Industrial Revolution, all value became synonymous with commercial value, and the spiritual, ecological, cultural and social significance of resources was eroded.

Vandana Shiva,
Water Wars: Privatization, Pollution and Profit

DOI: 10.1057/9781137062499

Contents

DOI: 10.1057/9781137062499

DOI: 10.1057/9781137062499

DOI: 10.1057/9781137062499

Preface

In addition to considerations about water rights, property rights, and political concerns, water management involves financial considerations. Some of the most common considerations for values in water management include agriculture, ranching, industrial use, municipal use, mining, energy, navigation, recreation, and various uses that affect natural environments. Water treatment and flood control are also important considerations. Ultimately, such values provide a decision-making framework for water managers. In other words, choices about use reflect societal values.

Water managers tend to have narrow views on what they consider to be the value of water, however. Not all water use is market driven; therefore, a comprehensive understanding of local community values associated with water can inform decision making by water managers. We use the term "water manager" to encompass not only the singular person assigned this community duty but also the many councils and institutions who make decisions regarding local water resources.

Regarding conventional values of water, an abundance of research exists; yet, for intangible aspects such as conservation for its own sake and spiritual connections, research involving value is limited. There are volumes of works that estimate the monetary values of water; however, intangible values are often overlooked. In general, few studies endeavor to estimate worth for water that encompasses more than a monetary value. Overall, there is not much attention in the literature to non-conventional

DOI: 10.1057/9781137062499

uses of water. The few sources that exist usually contend that, after basic needs are met, the utility, or pleasure, derived from water also lies in recreational uses and in the appreciation of nature that the conservation of water provides. The tourism industry benefits from this use of water, as do animals and humans who enjoy better water quality. Conventional economic analysis has had to make adjustments to take into consideration the "value" of non-use of water.

Our understanding of value in water is clouded, however, when we consider some atypical uses of this resource. Ultimately, conventional economic measurements fall short as an accurate tool to measure all the components of value when we include topics such as cultural conservation (conservation for its own sake) and religious or spiritual uses of water. In this book, we focus on non-monetary community values of water and the local satisfaction derived from them.

Conventional ways of valuing water work well and do a fairly good job dealing with the value of water in pollution and health. It is fairly easy to understand and measure when water becomes unusable or unhealthful because of a pollution problem. We also have a good understanding of the economic value of additional water made available through conservation. Yet there are other considerations in conservation that impact water uses.

Rarely are water policy demand projections made that utilize much more than civil engineering considerations and elasticity of demand based on value as conventionally defined by markets. An understanding of the non-market factors that impact conservation will help water managers make comprehensive decisions about water policy. We suggest here that these non-market community values should be a part of local water development and planning. In addition, we argue that it is the inclusion of both monetary and non-monetary values that justifies trust in the position of water manager.

In Chapter 1, we address tangible values of water. Here we see that even by conventional microeconomic standards, valuing water is a complex matter. Chapter 2 allows us to consider the ways in which conventional valuation of water is sanctioned. We look at the historical basis for monetary valuation of natural resources, the current global context surrounding the sanctioning of commodification, and the power structures that align from this paradigm.

In Chapter 3, we focus on the 2005 United Nations (UN) Millennium Ecosystem Assessment (MA). The MA is a broad study that was intended

DOI: 10.1057/9781137062499

to provide an account of ecosystems worldwide. Incorporated into this report is the importance of community values.

We reflect more deeply upon non-monetary values of water in Chapter 4. Within this chapter, we look at specific community and spiritual values. In doing so, we can further contemplate the complicated nature of valuing water.

Finally, Chapter 5 addresses the need for water managers to step up to the role with which they have been bestowed. We argue that water is a human right and that water managers are as agents of public trust. Within this chapter, we also address gender equality, and we provide some suggestions for future inclusion of community values.

The studies we reference throughout this book are rich as they relate to their fields. We believe that they legitimate the concept that non-monetary values should be taken into account. In addition, they provide a basis from which to extend the field of knowledge with regard to inclusion of community values. To date, however, there is little data that link the objectives of water managers with their practice. Our book fills that gap.

DOI: 10.1057/9781137062499

About the Authors

Kira Artemis Russo recently received her Ph.D. from the Department of Politics and International Affairs at Northern Arizona University. Her research focuses on environmental policy, particularly water resources. She received a bachelor's degree in Broadcast Communication from Northern Arizona University, a master's in History from California State University, Sacramento and a second master's in Political Science from Northern Arizona University. Her interests include being with her children, art, music, writing, and hiking.

Zachary A. Smith is a Regents' Professor of Political Science at Northern Arizona University. He received his bachelor's from California State University, Fullerton and master's and Ph.D. from the University of California, Santa Barbara. A consultant both nationally and internationally on natural resource and environmental matters, he is the author or editor of 27 books as well as numerous articles on environmental and natural resource policy topics. He currently teaches environmental and natural resource policy and administration in the public policy Ph.D. program at Northern Arizona University.

List of Abbreviations

CBA	cost-benefit analysis
CV	Contingent Valuation
EA	Environmental Assessment
EE	Ecological Economics
EPA	Environmental Protection Agency
GATT	General Agreement on Tariffs and Trade
GDP	gross domestic product
GSEGI	global state economic government institutions
IBRD	International Bank for Reconstruction and Development
IFI	international financial institution
IMF	International Monetary Fund
ITO	International Trade Organization
MA	Millennium Ecosystem Assessment
MIT	Massachusetts Institute of Technology
TK	traditional knowledge
UN	United Nations
UNESCO	UN Educational, Scientific and Cultural Organization
WTO	World Trade Organization

DOI: 10.1057/9781137062499

1
Conventional Values of Water

Abstract: *This chapter addresses tangible values of water. Here we see that even by conventional microeconomic standards, valuing water is a complex matter. In order to set the stage for a discussion of water marketing, we introduce the terminology that we will use throughout the book, and we briefly address the idea of commodification. After that, we begin the challenging discourse about value. We look at the literature regarding both conventional and non-conventional values of water, and we place these in the context of policy learning. Finally, we examine specific ways in which water managers place value on water in a market system, and we include various branches of economics that allow broader views on the topic.*

Keywords: Value of water, conventional value of water, non-conventional value of water, water management, monetary value of water

Russo, Kira Artemis and Smith, Zachary A. *What Water Is Worth: Overlooked Non-Economic Value in Water Resources*. New York: Palgrave Macmillan, 2013. DOI: 10.1057/9781137062499.

Introduction: the many facets of value

As a consumable commodity, water serves many uses. The most obvious of these are water for human consumption and water for food production. Other less apparent uses of water are those for mining, industry, municipalities, energy, navigation, and recreation. In addition, water is used to support the environment.

In this chapter, we address these tangible values of water. Here we see that even by conventional microeconomic standards, valuing water is a complex matter. In order to set the stage for a discussion of water marketing, we introduce the terminology that we will use throughout the book, and we briefly address the idea of commodification. After that, we begin the challenging discourse about value. We look at the literature regarding both on conventional and non-conventional values of water, and we place these in the context of policy learning. Finally, we examine specific ways in which water managers place value on water in a market system, and we include various branches of economics that allow broader views on the topic.

Terminology

In the context of this book, we use the word "commodification" rather than the term "commoditization." The latter emerged in 1965 from business theory.[1] This is the way in which goods become discernible among others in the marketplace. Unique qualities distinguish them from others so that the market can create a perfect competition among competing brands.

On the other hand, "commodification" originated in 1968 from a resurgence of Marxist theory to describe the process through which value is assigned. Karl Marx stated that a commodity derives its use-value from its ability to "serve the conveniences of human life."[2] He described exchange-value such that two commodities of unequal use-value can be reduced to a third non-commodified value that is different from the use-values of either of the commodities. In other words, two commodities of unequal value can be traded through the use of a third, substituted, value.[3]

Monetary values define the ways in which water is typically used and how it serves financial purposes. We use the word "monetary" to define values that can be commodified and, therefore, exchanged. Monetary values can be symbolized through units of currency. Once they are symbolized, these

DOI: 10.1057/9781137062499

values can be exchanged or traded in numeric ways, even in future time. Money is not only a means of exchange, but it also stores value.

As stated in the preface, the underlying assumption is that choices about use reflect human values. When market values become the substituted value for two commodities, two or more social values can be represented and exchanged. Differences between use-values and exchange-values often become apparent, however, when we address future substitutability. Accordingly, disconnect between real and substituted values of commodities often results in environmental problems.

We use the term "utility" as it is defined within the discipline of economics. As such, the word embraces the idea of pleasure or satisfaction derived from a good or service.[4] This definition often includes the more popularized notion of the term regarding utilization or use; here we contend, however, the criteria of pleasure or satisfaction might also denote non-use.

Within this paper, the word "economic" is defined by Paul Samuelson, a former professor at Massachusetts Institute of Technology (MIT) and a Nobel Laureate. He defines economics as "the study of how society ends up choosing, *with or without the use of money*, to employ scarce productive resources that could have alternative uses, now or in the future" (emphasis added).[5] Use of the word "economic" also denotes a distribution of wealth or an accumulation of goods or services. More broadly, the word is addressed here as ecosystem services as defined in the Millennium Ecosystem Assessment (MA). The MA describes ecosystem services as "the benefits people attain from ecosystems."[6] This definition includes all of the following uses of environmental resources: provisioning, regulating, supporting, and cultural uses.

The words "exchange" and "trade" are used interchangeably here. Both can be applied to the terms monetary as well as economic. Additionally, the words "societal" and "cultural" are interchanged in this paper.

Commodification

Marx asserted that money and value are not necessarily connected. "Money is the universal and self-constituted *value* of all things. It has therefore deprived the entire world—both the world of man and of nature—of its specific value."[7]

Substitution of money to exchange two or more values is, for that reason, subject to scrutiny. Commodification can serve to represent both

DOI: 10.1057/9781137062499

physical and moral values within society; paradoxically, this strategy can also undermine this worth. Igor Kopytoff explains how commodification represents societal values:

> From a cultural perspective, the production of commodities is also a cultural and cognitive process: [C]ommodities must not also be produced materially as things, but also culturally marked as a certain kind of thing. Out of the total range of things available in a society, only some of them are considered appropriate for marking as commodities. Moreover, the same thing may be treated as a commodity at one time and not at another.[8]

Kopytoff further states that shifts in differences about what is chosen to be commodified reveal a moral economy. We argue here that it is important for water managers to define when commodification represents principled social values and when it does not.

Highlights from the literature

Conventional values of water

This section focuses on scholarly research that addresses the many facets of value; we address the topic of valuation itself more explicitly in Chapter 3. Comparative estimates of value regarding utility and investments for production are the most common ways to address valuation in the water management literature.[9] Some discussions of the value of water as a commodity include extensive calculations for global estimates of this worth while others focus on market values in local economies. Estimates that attempt a value for water on a global basis tend to be weak because they fail to take into account local issues.

Even studies that focus on narrow issues often neglect the complexity by which the tangible values of water must be measured. In the report titled "Water Sector Policy Review and Strategy Formulation: A General Framework," the Food and Agriculture Organization (FAO) of the UN addresses this difficulty.

> Estimating the value of water is not easy because its value varies with quality, use, location and time. During dry periods of the year, or during droughts or during drought years, water values will be much higher than in other periods. Moreover, certain seasons or times of the year may also be important... because of critical water demands for crop growth, heating, cooling, industrial production or shipping.[10]

Contingent valuation (CV) is one method that has developed over the past 25 years for assessing non-use or non-market values of natural resources. CV is a favored topic of study among scholars who study market values of water. Although this method is an accepted standard for valuation of non-quantifiable considerations, water managers do not commonly incorporate non-monetary values into their decision making.

Another preferred subject of researchers who study water management is government regulation involved with water delivery, water purity and associated increased costs.[11] These scholars focus on market distortions that result from government interference in these markets. In her book *The Economic Value of Water*, D. C. Gibbons argues that flexibility is obstructed by the legal system; therefore, value is related to time, place, and administrative arrangements governing its use.[12]

The discussion above is not meant to be comprehensive. It is simply intended to highlight the many ways scholars address conventional values of water. The literature concerning unconventional values is far less dense. In the next sections, we highlight ways in which scholarly research relates to unconventional values of water.

Unconventional values of water

As we have stated, the breach in scholarly research is glaring as it regards intangible aspects of value such as conservation for its own sake and spiritual connections. Vast numbers of people worldwide value water in ways that reflect non-monetary (non-commodified) values, yet these considerations are rarely incorporated into water management policy. Below we specify many ways in which communities value water in non-conventional ways. We elaborate upon these values in Chapter 4.

Perhaps it is the wide range in community values of water that has limited scholarly study on the subject; multi-faceted values can be obscure and difficult to define. Through a detailed survey of water managers, one of the authors of this book addressed several varied non-monetary social values and compared them to the ones water managers say they hold. This study allowed a look at whether commodification might be the dominant paradigm for water managers worldwide and, if so, whether this paradigm even allows for water managers to take into account non-monetary community values.

DOI: 10.1057/9781137062499

Conservation values

Since the 1970s, certain communities have developed an ethic of conservation that is independent of water shortage or cost. The U.S. metropolitan areas of Phoenix and Tucson, Arizona stand as examples of this ethic. These communities are similar in demographics, climate, and topography; however, over the past 40 years, Tucson has consistently used approximately 25 to 50 percent less municipal water. It has a different ethic of conservation.[13] In his book *Desert Cities: The Environmental History of Phoenix and Tucson*, Michael F. Logan highlights these differences. He states that, even though both cities promoted conservation, it was the citizens of Tucson that embraced this ethic. He argues that historically people in Tucson were more aware of water scarcity because they knew they did not have as much water to spare.[14]

Social attitudes toward conservation can change over time—sometimes even quickly. During a prolonged drought in the late 1970s and the early 1980s, this happened in Santa Barbara, California. The city's water managers engaged in a comprehensive—sometimes severe—water conservation campaign. When the drought ended and conservation measures were no longer needed, the public did not return to its previous water use levels.[15] Attitudes toward water use clearly changed in significant and permanent ways, despite water cost or scarcity.

Some communities have stronger future orientations than others. For them, utility lies in the concern for the future or future generations. This concern is also an important part of the culture of conservation.

Spiritual values

With regard to spirituality, water is often considered a force of life, representing the presence of a higher power. From Egyptian to Balinese to Native American cultures, the concept is remarkably similar. According to Paula Abrams, there are two main reasons that water serves as a central place in the practices and beliefs of many religions. First, it cleanses and washes away impurities and pollutants. Second, water is considered a primary building block of life.[16]

Early Egyptian civilization congregated along the Nile, and all life and transport existed there. The river was considered life-giving in the physical realm; as such, it gained spiritual value, and the desire to include water as part of worship became woven into Egyptian society. Egyptians believed water was the fundamental element in creation. In

DOI: 10.1057/9781137062499

fact, it was the only element present in the beginning. This churning, chaotic water was called Nu. The Egyptians believed that it was out of Nu that all life began.[17] Annual floods played a large part in this "cosmic order." The Egyptians actually welcomed the floods. "They saw them not as nature out of control—let alone as nature in need of control—but as a gift to be treasured."[18]

Other early cultures also assigned significance to water in spiritual ways. Aphrodite was not only the Greek goddess of love, but she was also goddess of the sea.

In Aztec culture, Chalchiuhtlicue was worshiped as the goddess of running water.[19]

All over the world we see the spiritual importance of water: In France, a temple sacred to the goddess Sequana is located at the source of River Seine, and the Marne River got its name from Matrona, Divine Mother. The ancient name of the Thames River in England is Tamesa or Tamesis, denoting a river deity.[20]

Various religions share similar beliefs about water, no matter how diverse these faiths are. Within this book, we address both Eastern and Western spiritual practices; interestingly, they embrace similar spiritual values in relation to water.

The Eastern beliefs include Jainism, Hinduism, Chinese Traditional Religion (including Taoism), Buddhism, Shinto, and Sikhism. The Western religious beliefs include Judaism, Islam, Christianity, and other spiritual and more ecologically oriented thought as addressed by Zoroastrianism and the Bahá'í faith. We also discuss ecological-based spirituality as embraced by some Native American, African groups, and others that hold spiritual value for water.

These illustrations highlight the importance of considering these social values. By better understanding the satisfaction so many different cultures derive from water, water managers can better understand their own roles. In doing so, they can cultivate public trust in government management of water supplies.

Espeland's book *The Struggle for Water: Politics, Rationality, and Identity in the American Southwest* highlights the importance of the transformation of pleasure and satisfaction to dollar values. Termed "commensuration," this process has been used for centuries. Espeland states that understanding the forms of commensuration and the people who embrace it is critical for realizing its consequences.[21] Commensuration allows us to compare, as in cost-benefit analysis (CBA), two different entities,

DOI: 10.1057/9781137062499

yet its outcome can be both positive and negative.[22] When we look at commensuration as it relates to societal values, we often find that these values are immeasurable and that they are often not accurately reflected. The links between numeracy, pragmatism, and respect are not inherent. In the vein of political economy, Vandana Shiva develops this idea in a discussion of the social consequences of economic reductionism through commodification.[23]

Claudia Pahl-Wostl *et al.* provide a beneficial look at management practices and their relation to culture.

> Culture is advocated as crucial to understand barriers to the adoption of technologies and new management strategies and a successful exchange of experience ... This reflects a shift in emphasis from focusing on "hard" technology[-]based centralized approaches to a "soft" path in water management embracing participatory approaches and delivering diverse water services matched to the user's needs.[24]

"Soft" need not, however, be defined by indeterminate features. We suggest here that policy change is made possible through a clearer understanding of how water managers currently make decisions and whether this is consistent with the ways in which they want to represent their communities.

Policy learning

Over time, there have been several approaches that have attempted to correlate economics with culture. Within this study, we address these and focus specifically on the tenets of Ecological Economics (EE), which we believe has, to date, offered the widest framework for the inclusion of cultural or societal values.

Ecological Economics is based, to a great extent, upon the notion that the conventions of a market economy fall short when it comes to understanding human behavior.[25] Market economics allows us to measure, but it often does not allow us to deal with complex issues effectively.[26] We deal with the literature from this topic more fully at the end of Chapter 3.

Peter May suggests that changes in policy occur for various reasons, one of which is that social dynamics change. Echoing Paul A. Sabatier and Hank C. Jenkins-Smith, he argues that policy learning is at the heart of policy change.[27] Representation of social values lies in the

DOI: 10.1057/9781137062499

acknowledgement of the ways in which people value water and whether or not those values are taken fully into account. Responsibility within the field of water management is critical. Comparing social value with action allows a first look at the considerations and responsiveness of water managers.

Use values of water

The prolific literature regarding the market value of water provides many ways of determining "value," and the methods employed to determine value varies. Here we address value in water management in order to establish an appreciation of the complexity of value, even by established conventional monetary standards.

The value of water as a consumable commodity

Monetary terms

Volumes of books have been written about the diverse branches of the discipline of economics. Our objective here is neither to catalog this work nor to critique it. In order to look broadly at how water managers determine value, we address the contemporary offshoots of this discipline that influence decision making in water policy. With regard to valuation and water policy, paradox abounds.

At the outset, it is important to state that the market price of a commodity does not necessarily reflect its "organic value." Tom Holm *et al.* suggest that monetary worth is mechanistic in that the environment is worth a value equal to that which it produces or its actual value on the market. According to these authors, organic value emanates from relationships with nature that extend beyond productivity or cash value. Such values often derive meaning through associations with sacred places in the environment.[28]

Accordingly, society's value of natural resources is not always accurately reflected in price. As a consumable commodity, in particular, water has different values. As a human right, water has extremely high value from a mechanistic point of view. In fact, many would claim, in that context, that its value is immeasurable. In contrast, however, many economists argue that price should be simply an indicator of temporal supply and demand. Terry Anderson and Pamela Snyder contend that

DOI: 10.1057/9781137062499

a free market creates appropriate pricing and solves problems related to water use—including distribution and scarcity.[29] This is obviously one way to deal with the use and distribution of natural resources, yet it is only one of many approaches.

In the context of commodification, water managers approach the pricing of water in several ways. Here we do not attempt to address pricing specifically; rather, we provide a general outline of how water managers currently place value on water. This observation will later provide an opportunity to see how non-market community values of water might realistically be placed into a conventional market system. In addition, it allows identification of the parts of a market system that are detrimental to the acknowledgment of cultural values.

In general, this discourse assumes institutional bias. In other words, the underlying assumption is that people who are chosen as water managers support the dominant paradigms that are in place within the institutions. These paradigms include—but are not limited to—those involving commodification.

Both microeconomics and macroeconomics address both supply-side and demand-side market economies. Regardless of the differences involved with these sub-disciplines of economics, both employ the use of commodification to define the value of water. Supply-side economics is generally grounded in macroeconomics. This theory supports the idea that supply or output is the basis for growth. Implicit within this theory is the idea that lowering the barriers for production will, in turn, spawn large-scale growth.[30] Demand-side economics attempts to address the inefficiencies associated with private sector decision making. Keynesian economists argue that one role of government is to manage the effects of these inefficiencies.[31]

A. Billi *et al.* distinguish between valuation regarding the hydrological/engineering approach to water management and the monetary/institutional approach. These scholars contend that the hydrological/engineering model tends toward supply-side economics; in other words, these are valuations of water that take into account hydrology or engineering aspects of water management that address production. This approach focuses on reduction of seepage and expansion of infrastructure; therefore, addressing these concerns would, in theory, generate growth. On the contrary, monetary/institutional approaches toward water management deal with practices toward improving efficiency between costs and benefits.

> [The latter approach] uses optimization techniques under alternative institutional policies, in order to maximize the benefit of an allocation... This approach follows from the joint analysis of production and environmental costs, and of demand conditions, and is at the basis of the introduction of demand-management policies, such as cost recovery, environmental taxes, [and] water use permits tradable on special markets.[32]

These authors assert that complex hydrological modeling has emerged from the desire to better integrate monetary and institutional approaches. Generating complex formulas that represent social values is common within the field of economics. Ultimately, Billi, *et al.* believe that multi-objective approaches are necessary in order to overstep the boundaries that define value.[33] Again, these are simply ways in which commodification is addressed within the field of water policy.

The greatest concern of those examining value in water management research is the fact that the cost of water often does not reflect unique values. In many situations, water is treated as a free resource. Estimating the value of water is complicated in both market and non-market situations when consumers expect it to be free or cheap and when water suppliers worldwide bolster those beliefs by charging only for transportation and storage of water.[34]

Our understanding of water value is also confounded when there is rarely a charge that reflects opportunity cost, the value of the next desirable use (of water) that one must give up if the first option is chosen.[35] Diana Gibbons contends that there are ultimately few incentives for "affecting an efficient allocation among competing demands."[36]

In the next section, we further define value. Within the context of this discussion, it is easy to see in what ways commodification is incorporated into conventional values of water. The larger question is whether this incorporation serves community values. In other words, when defined in these often intricate ways, can commodification actually reflect societal values?

The utility of water

Monetary terms

This section is much more specific with regard to utility. This lends to a better understanding of the nature of commodification and how water managers make decisions. Incorporating the language of the discipline of economics allows us to more fully appreciate the complexity by

DOI: 10.1057/9781137062499

which people value water and in what ways communities might differ in their preferences. In addition to treating water as an economic good, something that satisfies a need or want, it is beneficial for water managers to understand the utility of water as it is related to community values.

Although our focus is on non-monetary values of water, it is important to state, up front, that some communities value certain industries. In fact, these are sometimes the most obvious ways in which social values are incorporated into the status quo. For example, some communities highly value water use for agriculture and for navigation that supports shipping. To underestimate these values that support industry would be impractical. Yet, it would also be naïve not to question whether these values are actually embedded in culture or whether they emerge through manipulation by leadership of the elites. This question would provide interesting future research.

Looking through the lens of utility, we see the similarities between discussing monetary or non-monetary values. As stated above, utility of water measures the amount of pleasure or satisfaction derived from it. This pleasure can come from both use and non-use values. For example, two people might derive the same amount of satisfaction in different ways. One person might enjoy using water to grow a garden while another enjoys watching a river flow downstream. Does anyone *not* enjoy a cool glass of water on a warm day? Table 1 lists definitions for terms related to monetary (conventional) values of water.

Conventional definitions of value state that utility of water is high in places where water is scarce. Marginal utility represents the pleasure or satisfaction gained or lost from an increase or decrease in the consumption of a commodity.[37] As in the case of flooding, extreme proliferation of additional water creates negative utility for water.

Marginal productivity of human-made capital is the term that allows us to understand how one more unit of one good (in this case, water) can create extra output or production.[38] On the topic of productivity, human-made capital is subject to depreciation, a decrease in value over time.[39] Elasticity shows how changing one monetary variable affects others; ultimately, this shows us responsiveness, or how closely related certain entities are.[40]

Depending on the intended use of water, sunk costs associated with particular water projects vary. Sunk costs are expenditures that have already been incurred and cannot be monetarily recovered.[41] Various

DOI: 10.1057/9781137062499

TABLE 1 *Terms for monetary values of water (in alphabetical order)*

Term	Definition
Contingent valuation	Based on hypothetical scenarios, a method to assess non-use (non-market) values of natural resources
Cost-benefit analysis	A method of assigning dollar values to non-monetary values so that they can be traded
Depreciation	A decrease in value over time
Ecological Economics	A discipline that suggests a more holistic approach regarding the relationship between economy and ecology
Elasticity	A measure of a variable's sensitivity to change as it relates to another variable
Environmental Economics	A sub-discipline of economics, this field applies financial theory to environmental problems
Marginal productivity	The amount of productivity gained or lost
Marginal utility	The amount of pleasure or satisfaction gained or lost from an increase or decrease in the consumption of a commodity
Pareto Optimal	A tenet of new welfare economics which holds that an economic situation is optimal when giving to one individual would not make another worse off
Sunk costs	Expenditures that have already been incurred
Sunk loss fallacy	The idea that misconceptions of costs can lead to poor decision making
Utility	Pleasure or satisfaction derived from a commodity
Utility discount rate	A measure of how much pleasure or satisfaction will be derived from saving a commodity for future use
Welfare Economics	A branch of economics that addresses social welfare

theories address sunk costs differently. Some theorists believe that poor decision making is often perpetuated by an aversion to change course after costs have already been incurred. Sunk loss fallacy strengthens this idea of loss in that even a misconception of costs can lead to poor decision making. Other theories support the idea of disregarding past costs and looking only at future costs and compensation. Community values, which often elude quantification, can get caught up in this debate, yet even social values that have previously been ignored can attach meaning to policy decisions.

Some communities show greater preference toward future orientations. In financial terms, a utility discount rate measures how much we value saving the commodity for future use. In other words, a low utility discount rate discounts (disregards) the future slightly, while a high utility discount rate discounts (disregards) the future heavily.

DOI: 10.1057/9781137062499

Community values that might emphasize saving water for future generations—the ones that express a low utility discount rate—are willing to forgo heavy use of the resource in the present. This understanding of pleasure or satisfaction as it relates to time can relate to both quality and quantity of water. The idea of quantity is easiest to understand in this regard. An illustration of a low discount rate toward the aim of quality would be a community that is unwilling to introduce many pollutants into its water. In order to accomplish this goal, it might forgo industrial or agricultural production so future generations would enjoy clean water.

In financial terms, money is most valuable in the present because of its potential to earn interest—its ability to gain value. Future utility is different. Although it is expressed as either a low or high utility discount rate, pleasure does not earn interest, so to speak. Based on decisions made in the present, community values toward a certain use might, on the whole, increase or decrease, but the value of the initial investment does not grow.

Welfare economics

Welfare economics is a branch of the discipline of economics that specifically addresses social welfare. It derives from the use of microeconomic techniques to analyze equity, justice, and altruism. The assumptions of this model are that utility can be measured, preferences are stable, additional consumption diminishes marginal utility, and all preferences can be compared. Neoclassical economist Arthur Cecil Pigou, a student of Alfred Marshall, stated that the constant shifting of economic goals makes it impossible to compare today's values with those of tomorrow. He called this an "inevitable shortcoming" of the discipline of economics.[42]

Pigou believed that policy decisions are imbued with monetary value, and he stated emphatically that welfare economics is not separable from non-monetary welfare. Above we differentiated between monetary and economic value. In context, we see that Pigou's use of the word "economic" is synonymous with the word "monetary."

> Economic [monetary] welfare will not serve for a *barometer* or *index* for total welfare... The real objection then is, not that economic welfare is a bad *index* of total welfare, but that an economic cause may affect non-economic welfare in ways that cancel its effect on economic welfare [italics in original].[43]

DOI: 10.1057/9781137062499

New welfare economics developed from neoclassical views. The later discipline holds roots in the theories of Vilfredo Pareto, John Hicks, Nicholas Kaldor, and T. Scitovsky. New welfare economics addresses social welfare in an ordinal way. This measurement is termed Pareto efficiency, the idea that a situation is optimal only if giving to one individual would not make another individual worse off.[44] Several theorists have created social welfare functions, formulas meant to address social equity.

An example of new welfare economics might include taking another hundred acre-feet of water from a stream bed. If this did not affect the in-stream value of providing a livable fish habitat, then the situation might not be Pareto optimal, but it could be Pareto efficient. In other words, the fish might not have an optimal environment, but it could be efficient enough to be habitable.

Environmental Economics

Environmental Economics has emerged from the discipline of economics. According to David Pearce of the U.S. Environmental Protection Agency (EPA), the field began in the 1960s and is now a major sub-discipline of economics. Environmental Economics applies economic theory to environmental problems. It combines certain work in welfare economics and growth with current views on political economy and conservation.

> The central tenets [of Environmental Economics] are that environmental problems have their roots in the failure of economic systems to maximize human well-being, that environmental quality matters for human well-being and for more traditionally oriented economic growth objectives, and that efficient policy can be achieved through incentive design.[45]

Conventional financial analysis has made adjustments to take into consideration the "value" of non-use of water. Elaborating on theories of efficiency within this field, the practices that have become most widely accepted in environmental policy are cost-benefit analysis and Contingent Valuation.[46]

Cost-benefit analysis

According to Nick Hanley and Clive L. Spash, the U.S. federal water agencies were among the first to use cost-benefit analysis. The Bureau of Reclamation and the U.S. Army Corps of Engineers used CBA to

DOI: 10.1057/9781137062499

compare water projects that related to flooding, irrigation, recreation, and conservation.[47] Through CBA, dollar values can be assigned to community values, yet it is challenging to separate the utility of water from the considerations associated with its future monetary costs or benefits. The problems associated with discounting become obvious in that context. Further, it is difficult to weigh opportunity costs amid discount rates that include the complexity of both monetary and non-monetary values.

John Bellamy Foster speaks to the complications of rational value-comparison as in CBA. He suggests that this method is lacking because it fails to take into account the idea that many values are plural; they matter in many ways to the same people as well as between different people.[48] "And equally clearly, it is not the case that in order to make practical judgments among these values for ourselves, we must buy into some single master-value which we can use as a measuring rod."[49] Many might argue, however, that the "master value" would be commodification.

Contingent valuation

Contingent valuation is a method that developed in the 1970s mainly to assess non-use (or non-market) values of natural resources.[50] Based on hypothetical scenarios, it involves asking people the amount of compensation they are willing to forgo for specific environmental services.

One of the most prominent CV cases concerns Glen Canyon Dam. "The valuation question of concern was how much recreational rafting was worth, compared to the market value of the peak-load power supply."[51] The study attempted to quantify the value of power generation versus rafting. It ultimately found considerable monetary values for rafting. The U.S. Congress formalized new river flows when it passed the Grand Canyon Protection Act of 1992.[52]

Despite the fact that CV has been used for a few high profile cases, water managers rarely use this method. Marc Willinger sums up the problem regarding the complexity of the measurement of non-use value:

> Non-use values are generally non-market values. To measure such values on a monetary scale, one has to design some kind of artificial market in which people can express a demand for not using an item. Therefore, there is no guarantee that the estimated value will not be influenced by the design of the artificial market. In other words, the true value cannot be observed unless one is able to give a non-controversial design of an artificial market.[53]

DOI: 10.1057/9781137062499

In the article "The Era of Management Is Over," Donald Ludwig states, "We need to reconsider both the notion that the natural sciences are objective and value-free and the corresponding idea that economics is value-free."[54] He debates the idea that science can solve all of our problems and asks whether monetary issues should be tied to trust in science. In addition, he asks whether experts are the only people who can help in creating beneficial policy choices.[55] These are valid criticisms with regard to the field of Environmental Economics.

Ecological Economics

Ecological Economics is a discipline that suggests a more holistic approach regarding the relationship between economy and ecology. This discipline is based, to a great extent, upon the notion that established market economics falls short when it comes to understanding human behavior.[56] Conventional methods allow us to measure, but often they do not allow us to deal with complex issues effectively.[57]

While EE incorporates the paradigms of the discipline of economics, the discipline stresses that a reliance on monetary measurements misleads governmental institutions to believe that they are assessing many environmental concerns, when really they are often addressing only limited interests. Herman Daly and Joshua Farley believe that environmental preservation would be the product of a balance between use of resources based on need rather than on the desire for financial growth.[58]

The call for an interdisciplinary approach in EE is meant to deal with issues involving greater complexity, but this approach also has its limitations. Conventional market economics allows us to narrow our focus of study, but it also forces us to stay within those boundaries. When addressing specific topics, this is not usually difficult. Yet, when we attempt to address more complex issues, this inclusion becomes more challenging. The dichotomy here is that the creation of policy is based largely upon conventional assessments; consequently, it is difficult to incorporate the "soft" notions of EE into policy values. We will further discuss EE as it relates to decisions within the field of water management.

The preceding discussion shows many ways in which economists engage in dialogues regarding the value of water. This list is simply meant to define the terms that help show how community values of water can

DOI: 10.1057/9781137062499

be incorporated into the already-existing framework for policy making. Placing monetary value on water is a complex matter; it means different things in different contexts. Even when water managers are limiting themselves to market values, often one value is not easy to compare to another.

Understanding value in water is further clouded when we consider some atypical uses or non-uses of water. Ultimately, conventional market measurements fall short as an accurate tool to measure all the components of value when we include topics such as local cultural conservation or community religious and spiritual uses of water.

In the next chapter, we elaborate on the ideas set forth here. The remainder of this book focuses upon non-conventional values of water. An understanding of these non-conventional values can lead water managers to an important perspective of the utility of water by consumers and can improve the ability of local water managers to make water policy that includes comprehensive community values.

Notes

1 Online Etymology Dictionary, "Commoditization, 2001–2012," accessed September 2, 2012, http://www.etymonline.com/index.php?allowed_in_frame=0&search=commodification&searchmode=none; parts of the preceding paragraph and other selected sections were taken in part from Kira Russo's dissertation.

2 Karl Marx, *Capital: A Critical Analysis of Capitalist Production*, vol. 1. (London: Lawrence & Wishart, Ltd, 2003), 44.

3 Ibid., 44–45.

4 Robert B. Ekelund and Robert D. Tollison, *Microeconomics*, 2nd edn (Glenview, IL: Scott, Foresman and Company, 1988), 137.

5 P.A. Samuelson. *Economics, an Introductory Analysis* (New York: McGraw-Hill, 1967).

6 Millennium Ecosystem Assessment (MA), "Ecosystems and Human Well-Being: Synthesis" (Washington: Island Press, 2005), 155.

7 Marx, 67.

8 Igor Kopytoff, "The Cultural Biography of Things: Commoditization as [P] rocess" in Appadurai, Arjun, ed., *The Social Life of Things: Commodities in a Cultural Perspective.* (Cambridge: Cambridge University Press, 1986), 211.

9 See e.g. Terry L. Anderson and Donald R. Leal, *Free Market Environmentalism* (San Francisco: Pacific Research Institute for Public Policy; Boulder: Westview Press, 1991); Anna Lindgren, "The Value of Water:

DOI: 10.1057/9781137062499

A Study of the Stampriet Aquifer in Namibia," *Department of Economics, UMEA University*. UMEA University (1999), http://www.econ.umu.se/MFS/annali.pdf (accessed January 13, 2005); Charles D. D. Howard and P. Eng, "The Economic Value of Water," *Conference: Mountains as Water Towers* (November 2003) http://www.cddhoward.com/docs/Economic%20Value%20of%20Water.pdf (accessed January 13, 2005).

10 "Water Sector Policy Review and Strategy Formulation: A General Framework," *Food and Agriculture Organization of the United Nations*. Food and Agriculture Organization of the United Nations, Rome, (1995), accessed November 30, 2004, http://www.fao.org/docrep/v7890e/V7890E00.htm#Contents.

11 Terry L. Anderson and Peter J. Hill, eds, *Water Marketing: The Next Generation*. (Lanham, MD: Rowman & Littlefield, 1997).

12 D. C. Gibbons. *The Economic Value of Water*. (Washington, DC: Resources for the Future, 1986), 1.

13 S. McKinnon. "Cities Push Water Conservation: Use Less without Sacrifice," *The Arizona Republic*, January 6, 2005. AzCentral.com, accessed January 25, 2005, http://www.azcentral.com/specials/special26/articles/0105conserve-main06.html.

14 Michael F. Logan. *Desert Cities: The Environmental History of Phoenix and Tucson* (Pittsburgh, PA: University of Pittsburgh Press, 2006).

15 C. Murillo. "Steelhead Water Wars," *Santa Barbara Independent* (Santa Barbara, CA), December 11–18, 2003, accessed January 31, 2005, http://www.independent.com/cover/cover.html.

16 P. Abrams, "Water in Religion," *The Water Page* (2000), Water Policy International Ltd, accessed November 30, 2004. http://www.thewaterpage.com/religion.htm.

17 R. Deurer, "Creation Mythology," *Egypt Art* (1997), accessed November 30, 2004, http://members.aol.com/egyptart/crea.html.

18 J. Rothfeder, *Every Drop for Sale: Our Desperate Battle Over Water in a World about to Run Out* (New York: Penguin Putnam, Inc., 2001), 30.

19 "Chalchiuhtlicue." *Encyclopædia Britannica* (2004) Encyclopædia Britannica Premium Service, accessed December 23, 2004, http://www.britannica.com/eb/article?tocId=9022262.

20 Vandana Shiva, *Water Wars: Privatization, Pollution, and Profit* (Cambridge: South End Press, 2002), 136.

21 Wendy Nelson Espeland, *The Struggle for Water: Politics, Rationality, and Identity in the American Southwest* (Chicago: University of Chicago Press, 1998), 11.

22 Ibid., 28.

23 Shiva, 27.

24 Claudia Pahl-Wostl, David Tàbara, Rene Bouwen, Marc Craps, Art Dewulf, Erik Mostert, Dagmar Ridder, and Tharsi Taillieu, "The Importance of

DOI: 10.1057/9781137062499

Social Learning and Culture for Sustainable Water Management," *Ecological Economics*, Vol. 64, Issue 3 (January 15 2008): 484, accessed August 26, 2012. http://dx.doi.org/10.1016/j.ecolecon.2007.08.007.

25 L. Venkatachalam, "Environmental Economics and Ecological Economics: Where They Can Converge?" *Ecological Economics*, Vol. 61 (2007): 550–558, 552.

26 M. Ruth, "A Quest for the Economics of Sustainability and the Sustainability of Economics," *Ecological Economics*, Vol. 56 (2006): 332–342, 335; C.S. Holling, Lance Gunderson, and Donald Ludwig, *Panarchy*, Island Press, Washington, D.C., (2002), chapter 1.

27 Peter J. May, "Policy Learning and Failure," *Journal of Public Policy*, Vol. 12, Issue 4 (Oct.–Dec., 1992), 332.

28 Tom Holm, Diane J. Pearson, and Ben Chavis, "Peoplehood: A Model for the Extension of Sovereignty in American Indian Studies," *University of Minnesota Press, Wicazo Sa Review*, Vol. 18, Issue 1, (Spring 2003): 7–24, http://www.jstor.org/stable/1409431.

29 Terry L. Anderson, and Pamela Snyder, *Water Markets: Priming the Invisible Pump*. (Washington, D.C.: CATO Institute, 1997), 104.

30 James D. Gwartney, "Supply-Side Economics," *Library of Economics*, (2008), accessed October 17, 2012, http://www.econlib.org/library/Enc/SupplySideEconomics.html.

31 Ekelund and Tollison, 64.

32 A. Billi, G. Canitano, and A. Quarto, "The Economics of Water Efficiency: A Review of Theories, Measurement Models and Integrated Models" (OPTIONS méderranneennes, Series B, n° 57), 227.

33 Ibid.

34 Zachary A. Smith. *The Environmental Policy Paradox*, 4th edn (Upper Saddle River, NJ: Prentice Hall, 2004), 134.

35 K. G. Frederick. "Preface," pp. vii–ix in Gibbons, Diana C. *The Economic Value of Water*. (Washington, D.C.: Resources for the Future, 1986), vii.

36 Gibbons, 1.

37 Ekelund and Tollison, 139.

38 Ibid., 298.

39 Ibid., 343.

40 Ibid., 108.

41 Ibid., 182.

42 Arthur C. Pigou, *The Economics of Welfare*, Library of Economics and Liberty (1932), accessed November 24, 2012, I.I. 4. http://www.econlib.org/library/NPDBooks/Pigou/pgEW1.html.

43 Ibid., 6.

44 Ekelund and Tollison, 139.

45 David Pearce, "Working Paper: An Intellectual History of Environmental Economics," *U.S. Environmental Protection Agency*, (November 1, 2002),

accessed November 25, 2012. http://yosemite.epa.gov/ee/epa/wpi.nsf/09133da7fb9a95db85256698006641d1/372307b3d147371585256fda004acb6b!OpenDocument.

46 Ibid.

47 Nick Hanley and Clive L. Spash, *Cost-Benefit Analysis and the Environment* (Brookfield, VT: Edward Elgar Publishing Company, 1993), 4.

48 John Bellamy Foster *The Vulnerable Planet A Short Economic History of the Environment*, (New York: Monthly Review Press, 1994), 235.

49 Ibid.

50 Mohammed H. I. Dore, and Timothy D. Mount, eds, *Global Environmental Economics: Equity and the Limits to Markets* (Malden, MA: Blackwell Publishers Ltd., 1999), 55.

51 R. C. Bishop, M. Welsh Brown, and K. Boyle, "Grand Canyon and Glen Canyon Dam Operations: An Economic Evaluation," in *W-133 Benefits and Costs in Natural Resources Planning, Interim Report #2*. Dept. of Agricultural and Resource Economics (University of Maine: Orono, 1989), http://www.ecosystemvaluation.org/contingent_valuation.htm#case3.

52 Ibid.

53 Dore and Mount, 55.

54 Donald Ludwig, "The Era of Management Is Over," *Ecosystems*, Vol. 4 (2001): 758–764, 759.

55 Ibid., 761.

56 L. Venkatachalam, "Environmental Economics and Ecological Economics: Where They Can Converge?" *Ecological Economics*, Vol. 61 (2007): 550–558, 552.

57 M. Ruth, "A Quest for the Economics of Sustainability and the Sustainability of Economics," *Ecological Economics*, Vol. 56 (2006): 332–342, 335.

58 H. Daly, and J. Farley, *Ecological Economics: Principles and Application* (Washington, D.C.: Island Press, 2004), xx.

DOI: 10.1057/9781137062499

2

Economic Grounds for Current Practices of Water Management

Abstract: *This chapter highlights the ways in which societies sanction the discipline of economics as a way to reflect value. We use the word "sanction" here to denote permission or approval from governmental authority that validates a course of action, or in this case, use. Seen in this light, the sanction of the discipline of market economics influences or dictates the formation of choice. Before we address water management and non-monetary values, it is important to look at the indoctrinations from which attitudes about market economics emanate. With these in mind, we can better understand why choices about monetary uses of water use are sanctioned. A glance at Western political theory provides the basis for understanding the economic grounds that guide current practices in water management.*

Keywords: Uses of water, water management, government sanction of use of water, Western political theory and water

Russo, Kira Artemis and Smith, Zachary A. *What Water Is Worth: Overlooked Non-Economic Value in Water Resources*. New York: Palgrave Macmillan, 2013.
DOI: 10.1057/9781137062499.

A starting point: values sanctioned in economics

In Chapter 1, we provided an overview of societal values of water and the utility derived from these values. In this chapter, we expand on that discourse. In essence, we highlight the ways in which societies sanction the discipline of economics as a way to reflect value. We use the word "sanction" here to denote permission or approval from governmental authority that validates a course of action, or in this case, use. Seen in this light, the sanction of the discipline of market economics influences or dictates the formation of choice.

Before we address water management and non-monetary values, it is important to look at the indoctrinations from which attitudes about market economics emanate. With these in mind, we can better understand why choices about monetary uses of water use are sanctioned. A glance at Western political theory provides the basis for understanding the economic grounds that guide current practices in water management. This discussion serves as the premise for the following questions: Do economic standards used in water management currently align water managers with community values? In addition, how can community values be incorporated into decisions about use?

First, we consider briefly the structural formation of a global economy and the resultant view of commodification—turning nature into capital. We focus on the views of Karl Marx and the transformation from use values to exchange or trade values. Next, we address hegemonic powers that result from the adoption of exchange values through a market system and associated institutions, and we look at why commodification becomes part of the social norm to water managers. Aristotle described this as the "common sense." In other words, why might water managers trade away hydrological services for capital? In order to strengthen our arguments about social relations and power, we supplement them with views from Karl August Wittofgel's "hydraulic society" and world-systems theory. In Chapter 5, we address the implications of the commodification of water.

Turning nature into capital

In Chapter 1, we addressed Marx's concepts of use value and exchange value. To reiterate, use values are inherent, or what Holm *et al.* refer to as organic values. Exchange values are values that can be traded through the use of a third value that is not a commodity itself and is different

DOI: 10.1057/9781137062499

from the use values of the other commodities.[1] According to John Bellamy Foster, Marx's work "contains numerous remarkable ecological insights."[2]

Until the past few centuries, human exploitation of the environment has existed on a fairly small scale. That situation changed with the emergence of the capitalist world system that began in the late 1400s.

The great historical transformation initiated by Columbus' voyage across the Atlantic five hundred years ago marked the origins of what was to become the capitalist world nation states, defined by the relation of colonizer and colonized, more developed and less developed.[3]

European colonization led to the extraction of substantial amounts of natural resources, which netted great financial gain. These resources were diverse: spices, sugar, coffee, tea, tobacco, furs, and precious metals. Later the colonizers brought additional sources of food back to Europe. Foster asserts that all these efforts helped set the stage for the next phase of capitalism, the Industrial Revolution.[4]

The Industrial Revolution led to the rapid increase in "the scale and intensity of production and to the development of a set of divisions that are at the core of our understanding of the modern world: the division between capital and labor, between economy and nature, and between center and periphery."[5] After the Industrial Revolution, use of natural resources contributed to new systems of large-scale production for capitalist profit. In many places worldwide, slave labor lowered production costs, thereby increasing the margin of profit for commodities. The examples of divisions between capital and labor are numerous and fill volumes in the annals of human history.

We focus here on the division between the economy and nature as it relates to water. Early colonists recognized through riparian rights the inherent value of water. "Early riparian principles were based on the notion of sharing and conserving a common water source. They were not attached to property rights."[6] Community values were often tied to usufructuary rights, rights that allow a person to use and enjoy the advantages of another person's property. Even colonists that settled in the eastern part of the U.S. conformed to this convention.[7] In essence, riparian rights relied on a "tort-like" system wherein privilege was tied to good behavior with regard to water use.[8]

Shiva defines the Doctrine of Prior Appropriation as the mechanism by which water was confined to a capitalist market. This doctrine emerged in the Western U.S., mainly to accommodate early hydraulic mining in

DOI: 10.1057/9781137062499

California and Colorado and early irrigation settlements in Colorado.[9] Defined as "first in time, first in right," prior appropriation provided absolute water rights to ownership of property. In so doing, users could compete for water, and they could develop far from the streams or rivers from which they laid claim.[10] Competition was the basis for the water market, and the later development of large dams provided the inventory.

Anderson and Snyder claim that the Doctrine of Prior Appropriation laid the foundations for water marketing. This doctrine created a system in which property rights were well-defined, enforced, and transferable.[11] Ultimately, these conditions provided for efficiency in a market, adding water to the many natural resources that could be commodified and traded. To embrace the perspective imbued in the Doctrine of Prior Appropriation, one must accept the idea that only a few people can own rights to water. Anderson and Snyder are not alarmed by this. They contend that even John Wesley Powell, one of the earliest explorers of the West, recognized this as truth.[12]

Experience taught early water developers that large-scale reclamation projects such as dams required massive amounts of capital. Developers had seen that extensive speculation of small companies had led those companies to financial failure. Nevertheless, population was expanding, and dams were considered the key to opening up the West.[13] "Gradually, a consensus built up in the West around [the] notion that the federal government should fund large-scale reclamation projects."[14]

At the turn of the century, the reclamation movement was successful in large part due to the emotional aspect that embraced the idea of the yeoman farmer as set forth by Thomas Jefferson. Mainly, reclamation of water resources would allow individuals to make land productive and would allow private ownership. These projects were sold based on the rhetoric of benefit to society.[15] In the next section, we discuss the theoretical foundations of power that underlie water marketing. In doing so, we approach whether water managers are, in fact, able to incorporate non-monetary values in decisions about water use.

Water markets and hegemonic powers

In this section, we explore the global permeation of commodification as the "common sense" for policy makers, including water managers. We look at classical liberalism and neoliberalism. In doing so, we lay

DOI: 10.1057/9781137062499

the groundwork for understanding how a market becomes entrenched through the institutions that have evolved around global trade, and we provide the foundation for comprehending that hegemonic powers have emerged largely through the creation of these institutions.

Here we use the word "institution" as it is defined by W. Richard Scott. Institutional theory highlights the ways in which associations guide human behavior.

> It considers the processes by which structures, including schemas, rules, norms, and routines, become established as authoritative guidelines for social behavior. It inquires into how these elements are created, diffused, adopted and adapted over space and time; and how they fall into decline and disuse.[16]

We address these structures in relation to the global market and the ways in which commodification has become the dominant paradigm for water management. We speak specifically to the institutions of property rights, commodification and organizations that support a market economy. In other words, we show how nature is used to support capital.

The indoctrination of classical liberalism

In general, classical liberalism advocates for civil liberties for the individual, for limited government, and for a free market.[17] Theorists such as Adam Smith, Thomas Malthus, and David Ricardo address certain elements of this ideology, yet most of the concepts involved with classical liberalism are tied together in the works of John Locke. Therefore, we focus solely on his work to highlight the early foundations for reliance on commodification and a global market system. We first address Locke and the institution of property rights because they are tied to the commodification of water. Next we look at Locke's views about money. Even though these views are tied more to a *laissez faire* system of government, they lay the basis for globalized markets.

Locke on property

Locke, who published his *Two Treatises of Government* in 1689, believed God gave "the earth and its fruits to mankind in common."[18] Although he defined these as common rights, he also believed that an individual had the right to property as a means of self-sustenance. According to Locke, this was God's way of providing for man's natural right to life.[19]

DOI: 10.1057/9781137062499

He also placed an emphasis on beneficial use, which instructed ways to compare uses.

Locke believed an individual was entitled to own land when he mixed his labor with it. The following statement sums up Locke's view of an individual's rights to own property that had been previously held in common:

> God gave the world to men in common; but since he gave it to them for their benefit, and the greatest [conveniences] of life they were able to draw from it, it cannot be supposed he meant it should always remain common and uncultivated. He gave it to the use of the industrious and rational (and *labour* was to be *his title* to it).[20]

Limitation of appropriation was born in this concept, meaning that an individual should be entitled only to the amount of land that he could work. Within the limitation of appropriation, Locke also incorporated the idea that the land, or its fruits, must not spoil, because God creates nothing so that it can simply be destroyed. In Locke's utilitarian views, the idea of preservation existed only to the extent that land had not yet been used.

Locke on money

In general, Locke's work supports "many of the facets of modern capitalist life," but he was, overall, apologetic for the emergent capitalist order.[21] As stated above, Locke believed in *laissez-faire* economics. He also believed in the sanctity of law and the liberty of individuals to exchange their property at market value. During his time, market value did not purport the use of a market economy; rather, it resided in the trade value of a commodity under the assumption that there should be minimal government interference. Huyler argues that the concept that Locke's views were capitalist emerge from a centrist view:

The Marxist or materialist-centered approach to Locke has been unfortunate and, in several ways, unproductive. By working backward, i.e., by working not from Locke's deepest commitments, but from our own modern conceptions (e.g. our view of "egoism," "individualism," "class," "capitalism," and even "exploitation"), we have largely missed the critical distinctions and nuances that informed Locke's political and economic outlook.[22]

For Locke, legitimate economic activities proceed from and depend upon moral commitments to God. Locke implied that the ultimate

DOI: 10.1057/9781137062499

interest lies not in wealth, but in the duty to that bond between man and "his Maker."[23] "What Locke presents is a theory of capitalism rooted not in class exploitation, but in certain intractable moral and metaphysical postulates that forbid the practice of political exploitation."[24]

According to Locke, man invented money, which could be used by consent in mutual exchange "for the truly useful, but perishable supports of life."[25] There is no need for money when land produces in abundance; the value of money lies in its ability to provide equal exchange for things that are disproportionately distributed in land.

Locke believed that the state of nature of man is that men seek to avoid pain and to receive pleasure. In order to counteract this individual attribute, men must unite into a community and "give up all the power necessary to the ends for which they unite into society, to the *majority* of the community" (original emphasis).[26] Ambiguity lies in the fact that Locke sometimes suggested law to uphold order, and he sometimes suggested that the government should use coercive power to protect rights. Nevertheless, Locke believed that there could be little unity among political interests if government did not rest upon the decisions of the majority. He contended that men must also hand over to society the jurisdiction over their possessions, including land. With that jurisdiction comes the protection of the individual's right to property.[27]

According to Locke, the role of government is to encourage production by providing appropriate protection of property, of interest rates, and of the value of money. Money is to be the measure of commerce, but its value should be set by the consent of the market, "not the variable and inconstant designs of the government."[28] Locke believed that a market economy should offer a favorable trade balance. A free market and government protection would shape national prosperity, which ultimately benefits society.

As we discuss the ways in which water use is sanctioned by governmental entities, we emphasize how deeply classical liberalism is entrenched in global decisions about trade. We argue that these views are now so intertwined into society that they are difficult to separate. Consent to these institutions is bound and displayed in rules about property rights, money and trade, and government protection of financial investment. Decisions about water are tied to property ownership, and the development of a market economy allows for trade of all natural resources, including water. Within this context, non-monetary uses are often overlooked or excluded.

DOI: 10.1057/9781137062499

The indoctrination of neoliberalism

Below we look at the ways in which the ideas of neoliberalism have become pervasive such that they would influence water managers worldwide. In order to distinguish between Locke's views about a market and exchange value, we look at the origin and meaning of the term "neoliberal." This discussion is based on the work of Taylor C. Boas and Jordan Gans-Morse, who have researched the term and the connotations of this ideology. After that, we address the permeation of these views throughout the global economy.

Although current scholarship identifies Milton Friedman and Friedrich Hayek as the fathers of neoliberalism, this school of thought, in fact, emerged from German economists. The term "neoliberalism" first appeared in writings about political economy between the two World Wars. The term had a different connotation than it does today.

Whereas contemporary scholars often equate neoliberalism with market fundamentalism, the Freiburg School's faith in the free market was moderate and pragmatic when compared to that of nineteenth century liberals. First, Freiburg School economists argued that for a free market to function, the state must play an active role.[29]

In fact, German neoliberals accepted the argument from classical liberal thought that powerful private actors might create threats to freedom of competition. "They argued that a *laissez faire* state policy stifles competition as the strong devour the weak" (italics added).[30] Alfred Müller-Armack coined the term "social market economy" to reflect the idea that social interests must be served. Boas and Gans-Morse clarify that these ideas were not necessarily meant to be altruistic; more likely they were meant to restore the social order following the previous political chaos. Hayek's work was tied intellectually to the German neoliberals, and he even contributed to one of their publications, but he was much more fervently opposed to state intervention in economic matters.[31]

Permeation of neoliberal views did not occur rapidly. For decades, use of the term was limited to Germany, but the concept gained acceptance in Latin America (mainly Chile), when pro-market intellectuals became aware of the views of the Freiberg School. In 1955, the University of Chicago pioneered a postgraduate program for Chilean economics students. These students, termed the "Chicago Boys," worked directly under Freidman, who was a professor at the University of Chicago. Although it was not yet called neoliberalism there, this philosophy was

DOI: 10.1057/9781137062499

dominant among Chilean right-wing intellectuals during the time of the 1973 coup that put Augusto Pinochet into power.

Though the first several years of Pinochet's military government were characterized by an ambiguous economic policy stance, the Chicago Boys eventually emerged as the primary architects of a reform program that sought a truly radical transformation of Chilean economy and society.[32]

Boas and Gans-Morse argue that neoliberalism is still a vague and contested term, but its worldwide transformation between the 1960s and the 1980s has largely encompassed Hayek and Friedman's views, which argue against state intervention. In the following paragraphs, we address neoliberal views as they relate to the development of the "common sense" of global trade and a hands-off approach to governance. Following that, we tie this global acceptance of market values to hegemonic power as it relates to water.

The globalization of trade

Like the term "neoliberalism," "globalization" is a contestable term. The complexity of the word lies in the relative newness of the phenomena that created it. These developments include rapid global travel, computer technology, multinational corporations, and interconnected world markets. According to William K. Tabb, it was not until the last decades of the twentieth century that awareness of globalization occurred. It became apparent that exports, investment, employment, and foreign direct investment were interconnected. It also became apparent that wealth and power had become more concentrated.[33]

After World War II, corporations became willing to venture into multinational markets; in doing so, they needed government protection. Coupled with emergent neoliberal views, this brought a new focus to the role of government involvement and international governance. We focus here on what Tabb refers to as the global state economic government institutions (GSEGIs) "of dominant importance."[34] These are the International Monetary Fund (IMF), the World Bank, and the World Trade Organization (WTO), which were organized to operate like nation states. "GSEGIs are instrumentalities of an evolving global governance system and are projections of power by the strongest states, most especially the United States."[35] Our purpose here is not to address specific policies of these institutions; rather, we seek to show the global

DOI: 10.1057/9781137062499

permeation of neoliberal ideas as displayed through the creation and continued endorsement of these institutions.

The International Monetary Fund

The IMF was conceived at the UN Monetary and Financial Conference at Bretton Woods, New Hampshire in July 1944.[36] Forty-four governments represented there sought to construct an institution that would "avoid a repetition of the vicious circle of competitive devaluations that had contributed to the Great Depression of the 1930s."[37] Negotiators at the Bretton Woods conference achieved a balance between domestic responsibilities of the state and a liberal world market.[38] Its original purpose was to promote monetary cooperation internationally. Tabb states, "This is no longer the case. The Fund now sees its job as forcing a restructuring of domestic financial policies in countries that go to it for loans."[39]

The World Bank

The World Bank was also devised at Bretton Woods. Originally known as the International Bank for Reconstruction and Development (IBRD), the bank was meant to assist with reconstruction after WWII. The promotion of foreign investment has always been central to its mission. "The rationale has been that without the confidence of private capital markets, development will not simply be difficult; it will be impossible."[40] Under the direction of former Secretary of Defense Robert McNamara, the World Bank underwent a "structural adjustment."[41] It liberalized financial markets, privatized some state enterprises, and when it could, forbade state interference into global markets.[42]

In the late 1990s, criticism of the World Bank was widespread, citing the idea that policies by international financial institutions (IFIs) were creating an indentured standing for low income countries. In effect, these countries were haunted with unending repayments for their debt. This recognition led to a split between the World Bank and the IMF. James D. Wolfensohn, President of the World Bank Group, addressed his institution's board of directors stating that it is imperative that the bank address social issues. He argued that without doing so there will be no political stability or financial stability.[43] Although many contest this as rhetoric, the organization considers poverty reduction and sustainable globalization its overarching goals.[44]

DOI: 10.1057/9781137062499

The World Trade Organization

Early intentions were to establish an institution for trade that complemented the two Bretton Woods organizations discussed above. Officials from more than 50 countries worked toward the end of creating an International Trade Organization (ITO) but "missed the flurry of support for internationalism that accompanied the end of WWII."[45] Between 1948 and 1994, world trade fell under the General Agreement on Tariffs and Trade (GATT) . According to Susan Ariel Aaronson, GATT was not a treaty, but more of a club. Its provisions were binding only as long as they were not inconsistent with a nation's existing legislation. GATT provided many of the rules that guided world trade; these rules accommodated high growth rates during its existence. The membership of GATT grew strongly from 1948 to 1993, but trade liberalization was costly for Americans. Workers there lost jobs due to increased foreign competition. By the late 1980s, the U.S. pushed to formalize GATT and give it broader powers. The World Trade Organization was the formalized version of that push. It was created on January 1, 1995.[46]

The WTO's powers now include many policies that were once the topic for national governments. Because many citizens worldwide see the institution as undemocratic, there have been several large protests against it. Criticisms against the organization center on the effect of trade rules upon human rights, labor rights, consumer protection, and the environment. Others argue that such agreements do not effectively regulate the behavior of global corporations.[47]

Permeation of the "common sense"

These three global institutions were created with a focus on export orientations over domestic use. Instilled within them are later neoliberal views, particularly those beliefs that markets should encourage trade and that there should be less government involvement, especially for corporations. Clearly commodification and market values are embodied as the dominant paradigm in these global institutions. General acceptance of these institutions defines the status quo—the "common sense."

In their book *Rules for the World*, Barnett and Finnemore address international organizations simply as bureaucracies. In doing so, they believe different expectations of such organizations are generated. "We see a world in which international organizations can act as good servants but can also produce undesirable and self-defeating outcomes."[48]

DOI: 10.1057/9781137062499

Theorists such as Tabb contend, however, that these outcomes are not accidental.

Tabb states that the global market was not created in a day. Complexity of this creation derives from the nature of institutions, from global to local, and guides this market.

> Accepting the free market outcome is not really a possibility because markets are always embedded in a larger societal framework. Markets need rules for contracting, standards for judging what is permissible behavior by participants, and enforcement of contracts.[49]

Tabb argues that the international institutions that deal with financial governance are enforcers of the political status quo. He contends that institutions such as the IMF and the WTO are guided primarily by capital. "While it enters from the outside, it is at the same time created through countless local adaptations to its dominance."[50]

The common sense of water marketing

Grounded in commodification, market economics is supported by dominant world institutions. Sometimes subtly and sometimes overtly, this view seeps into countless aspects of life. Our research reveals that the "common sense" of water marketing is pervasive. In the survey mentioned above, all water managers interviewed considered water as a commodity. We keep our minds open, however, to the idea that perhaps there are societies in which water marketing is not the "common sense."

Theorist Antonio Gramsci suggested that assimilation of views is not reserved to the intellectuals, but is capable of being imparted and displayed in hybrid ways. Further, he implies that feelings that form within the masses can be produced not only through formal education through a conscious leading group, but also through "everyday experience illuminated by 'common sense.'"[51] The permeation of commodification emanates not only from the world's dominant institutions, but it also proliferates in social interactions that reiterate this accepted view. Gramsci questions whether there is anything that can challenge the strong feelings of the masses when their views are established in such subtle ways.

Expanding on Marx, Gramsci argued that political and social preferences do not develop solely from economic struggle. These preferences also reflect assumptions about how society already exists and the normative aspect of how society should be. In a capitalist society, "these mediating assumptions are largely set by the ruling classes through their highly

DOI: 10.1057/9781137062499

developed mechanisms of political socialization."[52] Gramsci believed that once this sort of mystification occurs, the masses are powerless to overcome this moral and intellectual subordination.[53] Gramsci considers particularly powerful this predominance which is obtained by consent rather than the force of one class or group over other classes. He states that when domination is realized through the "coercive machinery of the state," moral and intellectual leadership becomes objectified in and exercised mainly through civil society.[54]

Because their positions fall in line with bureaucracy and associated formal institutions, the assumption here is that water managers represent the ruling class. In so doing, they impart the dominant paradigm of commodification and assent to the neoliberal views established by global and local institutions. As we stated above, bias is essentially instilled into policy through this position. The study mentioned above did not attempt to look at whether the bureaucratic nature of water management influences managers to overlook non-monetary values or whether their role is more dynamic within the realm of ruling class. Perhaps both conditions sway decisions by water managers on a situational basis.

Instrumentalism, the idea that science is useful for understanding, is particularly popular in the field of economics. In a chapter titled "The Modern Theory of Coloni[z]ation," Marx states, "Capital is not a thing, but a social relation between persons, established by the instrumentality of things."[55] As such, the tenets associated with it inherently contradict many values that are non-quantifiable.

Still, the question remains: Why might water managers trade away hydrological services for capital? The above-mentioned survey revealed that community values for water were included when they could be represented in some way by instrumental means. On the other hand, when they could not be represented this way, it was not the market that gave ground. It was community values that were less likely to be represented. In essence, water managers might not intentionally trade away hydrological services; rather, they might look to the dominant paradigm, which is inherently inconsistent with an ability to include values that are not quantifiable. In addition to some of the examples we provide in Chapter 4, future research in the field of cultural anthropology might reveal more values within certain communities that have been superseded by the pressure of the market.

Above we have looked at the formation of the value system—market economics—through colonization. In addition, we have considered the

DOI: 10.1057/9781137062499

creation and imparting of the dominant paradigm of commodification. As we stated above, a market system set in place the terms suitable for trade. Domination of the system occurred through property rights and institutional and legal arrangement that implied ownership. With regard to water marketing, the justification of the status quo was defined by the Doctrine of Prior Appropriation. The link between water and property became the means by which the status quo of market economics and commodification were justified.

In his book *Rivers of Empire: Water, Aridity and the Growth of the American West*, Donald Worster examines Wittfogel's "hydraulic society" theory. Wittfogel claimed that power of the elite is aligned, among other ways, with infrastructure for water. The more elaborate the infrastructure, the more likely power is to be held by the elite few. He argued that this power typically lies in a ruling class of bureaucrats. Wittfogel's work is based largely on that of Max Weber, the prominent sociologist who identified the importance of bureaucracy in relation to power. Shiva challenges Wittfogel's work, stating that small-scale bureaucracies have sometimes contributed to cooperation.[56]

Worster identifies different types of hydraulic communities: the local subsistence mode, the agrarian state mode, and the capitalist state mode. According to Worster, each of these exhibits increasingly powerful elites that manage these systems.[57] When we look at community values, the structure of power is important. Tabb states, "A hegemonic power often takes advantage of its strength to get what it wants and dresses its own interests as the general interest."[58] When these systems of competition are legitimized through the global market, arrangements of power become more apparent, especially in the ways that they lend to capitalist production.

Immanuel Wallerstein generally defines a world-system through division of labor. He explains further that geography dictates economic tasks and that these jobs are products of ecological considerations. "But for the most part, it is a function of the social organization of work, one which magnifies and legitimizes the ability of some groups within the system to exploit the labor of others, that is, to receive a larger share of the surplus."[59] Wallerstein and those researchers who embrace his theory claim that those countries at the core of the world economy are the ones at the core of production. Likewise, countries at the periphery of the economy have a peripheral focus on production.

Wallerstein states that a capitalist world economy "requires that groups pursue their economic interests within a single world market

DOI: 10.1057/9781137062499

while seeking to distort this market for their benefit by organizing to exert influence on states, some of which are far more powerful than others but none of which controls the world market in its entirety." Wallerstein's later work defines a core nation as one that holds dominance in all of the following ways: productivity dominance, trade dominance, and financial dominance. These nations are the most highly industrialized and the most diverse economically. In addition, Wallerstein states that core nations have stronger institutions that allow for the management of economic and monetary affairs; therefore, they hold significant influence over noncore nations. Based on changes in production, trade and financial capacity, core nations change over time.[60]

Within this chapter, we have provided the foundation for understanding how values of water can be sanctioned within the paradigm of commodification and the resultant market. In turn, we have considered the global institutions that support such a structure. In the following chapter, we look at the UN Millennium Ecosystem Assessment and globally shifting views on the inclusion of community values. In doing so, we begin to comprehend how important it is that water managers include these values in decisions about water use.

Notes

1 Karl Marx, *Capital: A Critical Analysis of Capitalist Production*, vol. 1. (London: Lawrence & Wishart, Ltd, 2003), 44–45.

2 John Bellamy Foster, *Marx's Ecology: Materialism and Nature* (New York: Monthly Review Press, 2000), 9.

3 John Bellamy Foster *The Vulnerable Planet A Short Economic History of the Environment*, (New York: Monthly Review Press, 1994), 13.

4 Ibid., 13–14.

5 Ibid.

6 Vandana Shiva, *Water Wars: Privatization, Pollution, and Profit* (Cambridge: South End Press, 2002), 21.

7 Ibid.

8 Jan G. Laitos and Joseph P. Tomain, *Energy and Natural Resources Law in a Nutshell* (St. Paul: WEST PUBLISHING CO., 1992), 357.

9 Dan A. Tarlock, "The Future of Prior Appropriation in the New West," *Natural Resources Journal*, Vol. 41 (2002), 770.

10 Shiva, 23.

11 Terry L. Anderson, and Pamela Snyder, *Water Markets: Priming the Invisible Pump*. (Washington, D.C.: CATO Institute, 1997), 104.

12 Ibid., 61.

13 Charles F. Wilkinson, *Crossing the Next Meridian: Land, Water, and the Future of the American West* (Washington, D.C.: Island Press, 1992), 243.

14 Ibid.

15 Ibid.

16 Richard W. Scott, "Approaching Adulthood: The Maturing of Institutional Theory," *Theory and Society*, Vol. 37, Issue 5 (October 2008), 428.

17 Robert Garner, Peter Ferdinand, and Stephanie Lawson, *Introduction to Politics* (New York: Oxford University Press, 2009), 60.

18 John Locke *Second Treatise of Government* (Indianapolis: Hackett Publishing Company, Inc., 1980), xvi.

19 Ibid.

20 Ibid., 21.

21 Jerome Huyler, *Locke in America: The Moral Philosophy of the Founding Era* (Lawrence: University of Kansas, 1995), 101.

22 Ibid., 102.

23 Ibid., 108.

24 Ibid., 164.

25 Locke, 28.

26 Locke, 53.

27 Ibid., xiii.

28 Huyler, 168.

29 Taylor C. Boas and Jordan Gans-Morse, "Neoliberalism: From New Liberal Philosophy to Anti-Liberal Slogan," *Studies in Comparative International Development*, Vol. 44, Issue 2 (June 2009), 146.

30 Ibid.

31 Ibid.

32 Ibid., 151.

33 William K. Tabb, *Economic Governance in the Age of Globalization* (New York: Columbia University Press, 2004), 39.

34 Ibid., 5.

35 Ibid.

36 "Bretton Woods Conference, 1944," U.S. Department of State, Office of the Historian (2013), accessed February 24, 2013. http://history.state.gov/milestones/1937–1945/BrettonWoods.

37 International Monetary Fund, "Factsheet: The IMF at a Glance" (2013), accessed February 24, 2013. http://www.imf.org/external/np/exr/facts/glance.htm.

38 Tabb, 109.

39 Ibid., 213.

DOI: 10.1057/9781137062499

40 Ibid., 195.
41 Ibid.
42 Ibid.
43 Ibid., 198–199.
44 World Bank, "History" (January 31, 2012), accessed February 24, 2013. http://web.worldbank.org/WBSITE/EXTERNAL/EXTABOUTUS/0,,contentMDK:20653660~menuPK:72312~pagePK:51123644~piPK:329829~theSitePK:29708,00.html.
45 Susan Ariel Aaronson. "From GATT to WTO: The Evolution of an Obscure Agency to One Perceived as Obstructing Democracy," Economic History Association (February 5, 2012), February 25, 2013. www.eh.net/encyclopedia/articel/aaronson.gatt.
46 Ibid.
47 Ibid.
48 Michael Barnett and Martha Finnemore, *Rules for the World* (Ithaca, NY: Cornell University Press, 2004), iv.
49 Tabb, 49.
50 Ibid., 59.
51 Antonio Gramsci, *Selections from the Prison Notebooks of Antonio Gramsci* (New York: International Publishers, 2005), 198–199.
52 Joseph V. Femia, *Gramsci's Political Thought: Hegemony, Consciousness, and the Revolutionary Process* (Oxford: Clarendon Press, 1981), 56.
53 Ibid.
54 Ibid., 24.
55 Marx, 717.
56 Shiva, 123.
57 Donald Worster, *Rivers of Empire: Water, Aridity and the Growth of the American West* (New York: Pantheon Books, 1985).
58 Tabb, 7.
59 Immanuel Wallerstein, *The Modern World-System: Capitalist Agriculture and the Origins of the European World-Economy in the Sixteenth Century* (New York: Academic Press, 1976), 230.
60 Immanuel Wallerstein, *The Modern World System II: Mercantilism and the Consolidation of the European World-Economy, 1600–1750* (New York: Academic Press, 1980).

DOI: 10.1057/9781137062499

3

The Millennium Ecosystem Assessment

Abstract: *This chapter addresses the substance of the United Nations Millennium Ecosystem Assessment, especially as it relates to non-monetary values and ecosystem and hydrologic services. First, we address the emerging global trend toward attending to community environmental values. Next, we attend to community values and ecosystem and hydrologic services. Finally, we recount the conclusions and progress following the MA. Overall, the conclusions of the report highlight the relationship between the desire to include community values and the ability to do so.*

Keywords: United Nations Millennium Ecosystem Assessment (MA), non-monetary values of water, ecosystem services, hydrologic services, community values of water

Russo, Kira Artemis and Smith, Zachary A. *What Water Is Worth: Overlooked Non-Economic Value in Water Resources.* New York: Palgrave Macmillan, 2013.
DOI: 10.1057/9781137062499.

An overview of the Millennium Ecosystem Assessment

In the year 2000, United Nations (UN) Secretary-General Kofi Annan called for the Millennium Ecosystem Assessment. The MA Sub-Global Assessments were meant to represent differentiated ecosystems worldwide. In general, they address current trends in ecosystem management, ecosystem services (benefits provided from the environment), and human well-being. The assessment is not the first to cite concerns about community values and the environment, but it is, however, one of the most comprehensive.

Because of its influence worldwide, this statement by the UN shows a far-reaching desire to include community and non-monetary values within environmental policy, including decisions about water. We include it here because it is generally representative of dominant environmental views and of accepted research methods. Kate A. Brauman *et al.* state that the Millennium Ecosystem Assessment is the formal international effort designed to elevate awareness and understanding of societal dependence on ecosystems and that it is the benchmark for ecosystem services research. As such, it illustrates the "wide-ranging importance of ecosystem services."[1]

In this section, we discuss the substance of this comprehensive report, especially as it relates to non-monetary values and ecosystem and hydrologic services. First, we address the emerging global trend toward attending to community environmental values. Next, we attend to community values and ecosystem and hydrologic services. Finally, we recount the conclusions and progress following the MA. Overall, the conclusions of the report highlight the relationship between the desire to include community values and the ability to do so.

The trend toward inclusion of community values

According to John McCormick, there is a misconception among those who study the environmental movement that the U.S. initiated and largely guides the pace of global environmental policies.[2] McCormick contends that the American environmental movement was influenced heavily by German forestry and that conservation was practiced in many parts of the world, including other parts of Europe, South Africa, and India. "It is difficult—even artificial—to assess environmentalism solely in terms of isolated national movements. To do so can lead to misconceptions."[3]

DOI: 10.1057/9781137062499

Nevertheless, one can see through an observation of U.S. environmental policy how policy makers have moved toward including community values in decision-making processes.

According to Daniel Fiorino, methods of environmental regulation that were implemented in the U.S. in the 1970s (especially those associated with command and control methods) were, at the time, appropriate responses by governments to industry, but these approaches became increasingly ineffective in dealing with new problems and social and economic relationships. He states that this sort of regulation was based on "fairly simple notions about industry's motivations and government's capacities."[4] Additionally, this type of management assumed that private industrial interests conflicted with the interests of society and that industry was seen as "amoral."[5]

This belief was not unfounded. Before this time, industry in the U.S. was responsible for widespread pollution that included the burning of Cleveland's Cuyahoga River and the oil platform rupture in the Santa Barbara Channel.[6] In addition, Lake Erie was considered a "dead" lake because mercury levels in the lake were too high to promote plant life. This depleted the amount of food and oxygen for fish. In her book *Global Spin: The Corporate Assault on Environmentalism*, Sharon Beder states that between 1965 and 1970 respect by the public for industry "fell to an all-time low" and the legitimacy of industry was called into question.[7]

Society also questioned whether government had the ability to determine and to decide what environmental goals should be and how those should be achieved.[8] The environmental regulation that followed was largely dependent upon bureaucracy and rational methods. Fiorino argues that this system encouraged heavy reliance on scientists, economists, attorneys, and administrators.[9] In other words, it relied strongly on expertise.

McCormick credits the 1968 United Nations Educational, Scientific and Cultural Organization (UNESCO) Biosphere Conference in Paris for creating an atmosphere conducive to further discussion on the topic of environmentalism. Although this global conference is minimally recognized, it set the pace for the 1972 UN General Assembly Stockholm conference. The difference between the two is that the Biosphere Conference looked specifically at scientific issues and associated environmental issues, while Stockholm was the first to address these in the context of "the wider political, social and economic questions."[10] Both perspectives broadened the discussion about environmental problems and created at the global level the forum from which the MA became prominent.

DOI: 10.1057/9781137062499

The conceptual framework for the MA situates "human well-being" as the focus for the assessment. This term is defined by situation, geography, ecological health, and culture.[11] In addition, the MA shows progress in surpassing what Fiorino refers to as the confrontational and intrusive environmental regulation of the past.

> It is impossible to devise effective environmental policy unless it is based on sound scientific information. While major advances in data collection have been made in many areas, large gaps in our knowledge remain. In particular, there has never been a comprehensive global assessment of the world's major ecosystems. The planned Millennium Ecosystem Assessment, a major international collaborative effort to map the health of our planet, is a response to this need.[12]

Although not its first purpose, the MA was also meant to help achieve the UN Millennium Development Goals. More than 1,360 social and natural scientists from approximately 95 countries contributed to the assessment.[13]

The assessment framework was meant to offer decision makers a tool for identifying ways to meet the demands for clean water, food, health, and employment. In addition it would help provide knowledge for improved decision making and build that capacity. Finally, it was intended to analyze and pass along this information.[14] In theory, this information would allow decision makers more options to meet growing population demands, to understand more fully who the stakeholders are and what the trade-offs might be. Ultimately, the report was meant to align this understanding with the appropriate governances that can affect change.[15]

Ecosystem and hydrologic services

The MA links ecosystem and hydrologic services and human well-being in several ways, involving both conventional and non-conventional values. In doing so, it speaks to the complex nature of such an investigation: "Well-being is multidimensional and so very hard to measure."[16]

Ecosystem services

The MA lists monetary valuation and health indicators as the main ways in which human well-being can be linked to ecosystem services, and it

DOI: 10.1057/9781137062499

notes literacy and mortality as other indicators that link these matters.[17] Terms of monetary valuation in the MA are used mainly to assess trade-offs between substitutable ecosystem management arrangements that change the use of ecosystems and, equally important, the services available by their use.

Monetary valuation is further applied in order to obtain a total value for an ecosystem or to increase that value. Poverty and equity are also defined in the MA through financial means. The valuation methods suggested in the MA parallel the ones we discussed in Chapter 2. Generally the report cites gross domestic product (GDP) and the Human Poverty Index as aggregate indicators of human well-being.[18] The MA states that overall monetary valuation endeavors to measure utilitarian benefits that ecosystems provide.

The report concomitantly acknowledges ecological, sociocultural, and intrinsic (organic) values of ecosystems, species, and even inanimate objects.[19]

> What might be called a sociocultural value derives from the value people place on elements in their environment based on different worldviews or conceptions of nature and society that are ethical, religious, cultural, and philosophical... These values are expressed through, for example, designation of sacred species or places, development of social rules concerning ecosystem use... and inspirational experiences.[20]

The dichotomy between recognition of monetary and non-monetary values in the MA is apparent; in fact, acknowledgement by the MA for non-monetary values does not necessarily ensure that they can be included in decision making. Indeed, if it is true that hegemonic powers influence the dominant paradigm involving commodification, then it is possible that decisions about ecosystems, including decisions about water, might be affected by that paradigm rather than by intrinsic or organic values. Further, prominence of this paradigm might reveal itself through the domination of one class over another.

The Conceptual Framework of the MA highlights the links between provisioning, regulating, and cultural ecosystem services. It underscores the relationship between sustenance of freshwater supplies and culture. Although it is not fully inclusive, the MA identifies certain aspects of culture as aesthetic, spiritual, educational, and recreational.

The authors of the report list all aspects of culture as links to "security, basic materials for a good life, health and good social relations."[21]

DOI: 10.1057/9781137062499

Ultimately, the language of the MA is grounded in the liberal view of "freedom of choice and action" and the "opportunity to be able to achieve what an individual values doing and being."[22]

The limitations of such an assessment are obviously vast. Considering human nature, culture cannot reasonably be separated from provisioning and regulating, nor can security be seen as distinct from having adequate access to basic materials. The definitive leap of the assessment to human rights can also be considered problematic if one contemplates the many variables associated with the opportunity to do and be what we value.

Because culture is considered a weak link within ecosystem services, one might argue that it is, therefore, relatively unimportant. That would imply, however, that just because it is difficult to explain, it does not matter. We contend, however, that the inclusion of this aspect in such a notable study as the MA stresses its relevance to water management. We further assert that cultural relevance might perhaps be minimized within this assessment in light of current political and financial paradigms that support the status quo involving commodification.

Hydrologic services

Concerning hydrologic services, the MA states that there is a growing dependence by human populations on these services. There have been many attempts to stabilize the ability to deliver water supplies, but this sort of influence by water managers has often been negative. "The capacity of ecosystems to sustain freshwater provisioning services is thus strongly compromised throughout much of the world and may continue to remain so if historic patterns of managed use persist."[23]

Neil S. Grigg states that the philosophical foundation of accepted water management revolves around the notion of sustainability, yet actions to achieve it bring forth conflict because they necessarily involve property, jobs, rights, and taxes.[24]

> Managing water resources requires skills and approaches that go beyond pure engineering, science, management, or law. In the [T]wenty-first [C]entury, water managers will deal with complexity and conflict. They will have to confront this complexity by analysis that enables them to unravel interdependency of systems, and they will have to confront conflict with cooperation, coordination, and communication, especially with the public.[25]

DOI: 10.1057/9781137062499

This statement illustrates the multi-faceted nature of water resources management, and it provides an understanding that community values for resource management are integral.

Approximately 2.6 billion people in the world (about a third) lack facilities for proper disposal of human waste, and 2 billion people face chronic water shortages. Additionally, nearly 1.1 billion people in the world still do not have a safe supply of clean water. Each year approximately 1.7 million people die as a result of poor sanitation; this equates to a loss of about 50 million healthy life years.[26] Of these, approximately 1.4 million (about 70 percent) are children under the age of five.[27]

These principal concerns for water management affect a considerable portion of the world population and are globally far-reaching. In addition, they make clear the responsibility water managers possess. Obviously, water managers serve as important representatives of public trust. We assert that this trust is further shaped through water management that considers community non-market values of water.

Conclusions and progress

In March 2005, the MA Board released a document outlining ten key messages of the study. Among the unresolved problems, it identifies the loss of ecosystem services, including water supplies. The board states that human impact on ecosystems is unlikely to be reduced until the full value of ecosystem services is taken into account.[28] In this section, we highlight the conclusions and progress in the MA regarding the inclusion of non-monetary values of water, especially those involving community values.

Among its conclusions, the MA asserts that communities frequently lack the ability to intervene when they become subject to socioeconomic powers that result in changes in the ecosystem. Researchers such as Elinor Ostrom have suggested nesting institutions and organizations to buffer such forces. The MA further identifies consideration of cross-scale interactions and social networks as key.

> These large-scale processes include policies, conventions, funding programs, market forces, tourism, global warming, and mega projects such as large dams and transboundary protected areas. Some have negative impacts; others can be used by communities to improve their well-being.[29]

The board contends that inclusion of community values can be beneficial to ecosystems and the services they provide. "Measures to conserve

DOI: 10.1057/9781137062499

natural resources are more likely to succeed if local communities are given ownership of them, share the benefits, and are involved in decisions."[30] Full value of ecosystem services would, theoretically, include both monetary values and non-monetary values.

The MA states that local management also plays a central role in generating ecosystem services both locally and globally. At times, community-based management has, in fact, created local changes to lessen the impacts of crisis, such as droughts or floods. At other times, however, community management systems are "dysfunctional."[31] Our study does not attempt a count of successful or dysfunctional systems of management. Rather, we strive to address whether non-monetary community values might be neglected if the paradigm of commodification holds influence over policy makers.

According to the MA, community-based ecosystem management, including traditional knowledge (TK), promotes resilience when full values are taken into account.[32] "Diverse cultures perceive this relationship in different ways, and institutionalize various rules of behavior (taboos) with regard to the sacred space and its elements."[33] These rules of behavior often include conservation practices and support ecosystems at the species level.

Two examples in the MA illustrate this point. According to Carl Folke *et al.*, locals in Western China have created elaborate underground water harvesting structures called *karez*. These 800-year-old traditional structures maintain both water quality and quantity. The next example involves people in the Laguna Lake Basin in Southeast Asia. They know the most lucrative fishing areas and where to place rice paddies with respect to local water flow patterns.[34]

> The community-based assessments illustrate that much more than ecological knowledge and understanding is necessary for strengthening the capacity to manage ecosystems sustainably for human well-being. The social dimension behind management needs to be accounted for as well.[35]

Regarding ways in which to incorporate multiple knowledge systems with the dominant paradigms involving science and commodification, the MA cites even implementation of the study itself as challenging.

> Particularly important were the lack of local and technical expertise to conduct the assessment and to deal with conflicting knowledge, and lack of familiarity with appropriate tools and methods with which to study local knowledge systems.[36]

DOI: 10.1057/9781137062499

In fact, our study may not reach far enough. "The MA worldview may exclude key alternative local framings, such as a dynamic landscape perspective that views biological patterns throughout the region as being shaped through the interaction of social and ecological processes over time."[37] Ultimately, these assessments present lessons for sustainability, and local groups and actors should be involved with collaboration, even though this is not easy to do. Long-term research makes possible shared knowledge for people involved.

A further look at Ecological Economics

Much of the MA references scholars who study Ecological Economics. In this section, we look further to this discipline to define ways in which it is possible to critically assess whether non-monetary values can truly be considered by water managers. In other words, are inclusions about non-monetary values only nominal, or are they genuinely reflected in policy? In evaluations such as the MA, this delineation is critical if we are to deal with current financial imperatives and the ability to include community values.

Although it is often vague, the link between science and the discipline of economics holds importance in this capacity. EE endeavors to come to terms with frequent dichotomous roles of science. Science is often necessary in order for us to comprehend the magnitude of resource depletion; accordingly, these observations often support commodification. If it is true that science often supports the discipline of economics, then questions about inherent or organic value remain.

The MA employs conventional data sources and methods, and its underpinning is the availability of data to monitor ecosystems. Remote sensing technologies and analytical tools such as geographic information systems "allow data on the physical, biological, and socioeconomic characteristics of ecosystems to be assembled and interpreted in a spatial framework, making it feasible to establish linkages between drivers of change and trends in ecosystem services."[38] In addition, future assessments will be based on this systematic data collection.[39]

Critics argue that the roles of science and the discipline of economics have become too prominent and that experts need to find ways to bridge the gap between specific and systemic solutions. Many proponents of EE argue for an ecologically based science that encompasses more expansive knowledge.

DOI: 10.1057/9781137062499

T.C. Brown asserts that conventional market economics lacks the ability to truly measure value.[40] In addition, this author argues that the ability to measure is based on conceptual and relational roles.[41] If, in fact, these relational roles become political, then Gramsci's theory of Cultural Hegemony seems reasonable—that a dominant class could manipulate the system of value and impart the paradigm of commodification that justifies its domination over other classes.[42]

One reason water managers might want to concern themselves with the possibility of hegemonic influence lies in the concept of trade value, which is listed as a prominent concern in the MA and other reports of this genre. If ecological degradation occurs, sometimes substitutions of knowledge and manufactured human capital can mitigate the depletion and degradation of ecosystem services. In other words, a market economy sometimes allows pardon in places where it is deemed that humans have damaged the environment.

The MA addresses the boundaries of the leniency in trade that a market economy provides through the equalization of value expressed in commodification. It specifically addresses the limits to substitution, especially with regulating, supporting, and cultural services. "For some people, especially the poorest, substitutes and choices are very limited. For those who are better off, substitution may be possible through trade, investment, and technology."[43] In many cases, however, there is no substitute for water.

The MA states, however, that no substitution is possible when we consider the cultural importance of the loss of an important species (such as whales and tigers). The report additionally acknowledges that market mechanisms do not ensure conservation of ecosystem services. This is especially true for cultural services, which typically do not have markets. Further, ecosystem services might not benefit people living within the ecosystem; rather, they might provide gain to others who are far away.

Regarding intrinsic or organic value, the MA stresses that something can serve as valuable for one person, but not for another.

> From the perspective of many ethical, religious, and cultural points of view, ecosystems may have intrinsic value, independent of their contribution to human well-being. The utilitarian and non-utilitarian value paradigms overlap and interact in many ways, but they use different metrics, with no common denominator, and cannot usually be aggregated, although both paradigms of value are used in decision-making processes.[44]

DOI: 10.1057/9781137062499

In this chapter, we have provided an illustration of how the global governance has attempted to unite community values with ecology. The question we continue to ask is whether this view is actually informing policy. In the following chapter, we address specific values of water that people hold worldwide. In doing so, we see how pervasive and far-reaching these values are.

Notes

1 Kate A. Brauman, Gretchen C. Daily, Duarte T. Ka'eo, and Harold A. Mooney, "The Nature and Value of Ecosystem Services: An Overview Highlighting Hydrologic Services," *Annual Review of Environment and Resources*, Vol. 32, Issue 1, (2007): 67–98, 62.

2 John McCormick, *Reclaiming Paradise: The Global Environmental Movement* (Bloomington, IN: Indiana University Press, 1991), viii.

3 Ibid.

4 Daniel Fiorino, *The New Environmental Regulation* (Cambridge, MA: MIT Press, 2006), 6.

5 Ibid.

6 Zachary A. Smith. *The Environmental Policy Paradox*, 4th edn (Upper Saddle River, NJ: Prentice Hall, 2004), 17.

7 Sharon Beder, *Global Spin: The Corporate Assault on Environmentalism* (White River Junction, VT: Chelsea Green Publishing Company, 1998), 15.

8 Fiorino, 6.

9 Ibid.

10 McCormick, 88.

11 Ruth DeFries, Stefano Pagiola, W. L. Adamowicz, H. Resit, Akçakaya, Augustin Arcenas, Suresh Babu, Deborah Balk, Ulisses Confalonieri, Wolfgang Cramer, Fander Falconı´, Steffen Fritz, Rhys Green, Edgar Gutiérrez-Espeleta, Kirk Hamilton, Racine Kane, John Latham, Emily Matthews, Taylor Ricketts, Tian Xiang Yue Augustin, Neville Ash, and Jillian Thönell, "Analytical Approaches for Assessing Ecosystem Condition and Human Well-being," Millennium Ecosystem Assessment (Washington: Island Press, 2005), 54, http://www.unep.org/maweb/documents/document.271.aspx.pdf, accessed January 10, 2012.

12 Kofi Annan quoted in "MA Conceptual Framework," Millennium Ecosystem Assessment. (Washington: Island Press, 2005), 26, accessed January 10, 2012, http://www.unep.org/maweb/documents/document.765.aspx.pdf.

13 "20 Projects to Showcase 20 Historic Years of Environmental Finance," Millennium Ecosystem Assessment Online Brochure. United Nations

DOI: 10.1057/9781137062499

Environment Programme (UNEP), accessed January 10, 2012, http://www.unep.org/dgef/Portals/43/news/facts/GEF%20Folder%20Inserts_12.pdf.

14 DeFries *et al.*, 26.

15 Ibid.

16 Ibid., 54.

17 Ibid.

18 Ibid., 62.

19 Ibid.

20 Ibid.

21 "Linkages between Ecosystem Services and Human Well-being" from "Ecosystems and Human Well-being: Opportunities and Challenges for Business and Industry," Millennium Ecosystem Assessment. (Washington: Island Press, 2005), 4, accessed January 10, 2012, http://www.unep.org/maweb/documents/document.353.aspx.pdf.

22 Ibid.

23 Charles J. Vörösmarty, Christian Lévêque, Carmen Revenga, Robert Bos, Chris Caudill, John Chilton, Ellen M. Douglas, Michel Meybeck, Daniel Prager, Patricia Balvanera, Sabrina Barker, Manuel Maas, Christer Nilsson, Taikan Oki, and Cathy A. Reidy. "Fresh Water," Millennium Ecosystem Assessment (Washington: Island Press, 2005), 168, accessed January 10, 2012, http://www.webpages.uidaho.edu/uiferl/pdf%20reports/MA%20Freshwater%20Ecosystem%20Services.pdf.

24 Neil S. Grigg, *Water Resources Management: Principles, Regulations, and Cases* (San Francisco: McGraw-Hill, 1996), 7–8.

25 Ibid., 4.

26 Vörösmarty *et al.*, 167.

27 "Estimating the Cost of Disease," *World Health Organization*, World Health Organization (2004), accessed June 28, 2004. http://www.who.int/peh/burden/articleEHP052002.pdf.

28 "Living Beyond Our Means—Natural Assets and Human Well-being: Statement from the Board," Board of the Millennium Ecosystem Assessment. World Resources Institute (March 2005), 3, accessed January 10, 2012, http://www.wri.org/publication/millennium-ecosystem-assessment-living-beyond-our-means-natural-assets-and-human-we.

29 Ibid.

30 Ibid.

31 Carl Folke, Christo Fabricius, Georgina Cundill, Lisen Schultz, Cibele Queiroz, Yogesh Gokhale, Andrés Marín, Esther Camac-Ramirez, Shivani Chandola, Mohamed Tawfic Ahmed, Bibhab Talukdar, Alejandro Argumedo, and Fabricio Carbonell Torres. "Communities, Ecosystems, and Livelihoods." Millennium Ecosystem Assessment. (Washington: Island Press,

DOI: 10.1057/9781137062499

2005), 275, accessed January 10, 2012, http://www.unep.org/maweb/documents/document.349.aspx.pdf.

32 Ibid., 273.

33 K. C. Malhotra, Y. Gokhale, S. Chatterjee, and S. Srivastava, *Cultural and Ecological Dimensions of Sacred Groves in India* (Indian National Science Academy, 2001), 1.

34 Folke *et al.*, 273.

35 Ibid., 275.

36 Polly Ericksen, Ellen Woodley, Georgina Cundill, Walter V. Reid, Luís Vicente, Ciara Raudsepp-Hearne, Jane Mogina, and Per Olsson, "Using Multiple Knowledge Systems: Benefits and Challenges," Millennium Ecosystem Assessment (Washington: Island Press, 2005), 114, accessed January 10, 2012. http://www.unep.org/maweb/documents/document.343.aspx.pdf.

37 Ibid.

38 DeFries *et al.*, 39–40.

39 Ibid.

40 T. C. Brown, "The Concept of Value in Resource Allocation," *Land Economics* Vol. 60 (1984), 244.

41 Ibid., 232–233.

42 Antonio Gramsci, *Selections from the Prison Notebooks of Antonio Gramsci* (New York: International Publishers, 2005), 199.

43 DeFries *et al.*, 32.

44 Ibid., 34.

DOI: 10.1057/9781137062499

4

Non-Conventional Community Values of Water

Abstract: *This chapter attends to the significance of water as it relates to spirituality, highlighting the meaning that various cultures assign to water and rites for which this natural resource is used. It is apparent that a majority of people worldwide place utmost importance on the organic or inherent value of water as it relates to spirituality. We argue that this is a value that should be incorporated into water development planning. The Eastern beliefs we include are Jainism, Hinduism, Chinese traditional religion (including Taoism), Buddhism, and Shinto. Regarding Western spiritual beliefs, we reflect on Judaism, Islam, Christianity, and other spiritual and ecologically oriented thought as addressed by Zoroastrianism and the Bahá'í faith. We also discuss ecological-based spirituality as embraced by some Native American and African groups.*

Keywords: Cultural value of water, spiritual value of water, organic value of water, inherent value of water

Russo, Kira Artemis and Smith, Zachary A. *What Water Is Worth: Overlooked Non-Economic Value in Water Resources*. New York: Palgrave Macmillan, 2013.
DOI: 10.1057/9781137062499.

Non-monetary values of water

The above discussion about conventional values of water shows the complexity by which water managers must address monetary values of water and its consequent use. In Chapter 1, we provided a general look at community spiritual values of water worldwide. Within this chapter, we address the topic further.

Attending to the idea of non-monetary value through conventional methodologies is even more complicated. Within this chapter, we address environmental and inherent or organic values of water as described above by Holm *et al.*[1] These illustrations allow a broader view of whether policy does, in fact, incorporate such community values into decision making. In this chapter, we broadly address meaning for these community values.

Environmental values of water

The acknowledgment of non-monetary values over the past few decades by policy makers is noteworthy. As we discussed in Chapter 1, cost-benefit analysis and contingent valuation work well to define value in many environmental situations. Nevertheless, these methods possess shortcomings that do not fully embrace values that communities hold.

In-stream values of water

Many in-stream values can be addressed through conventional methods. Values that include water for fish and habitats for other animals are fairly easy to account for. These are simply included as measurements of the quantity of water needed to support these environments. Overall, much of this value can be represented through the tenets of ecologically sustainable water management. Even certain spiritual values that require preservation through quantification of water can be represented this way.

Kira Russo's survey of a water manager in British Columbia revealed that certain uses for First Nations people in that area could easily be factored into decision making. If water were to be used for a bathing pool or for ceremonial purposes, then in general, legal frameworks support the ability of water managers to quantify that amount. More complex, however, is spirituality that permeates a body of water. In one instance, First Nations there claimed that roadwork would encroach

DOI: 10.1057/9781137062499

upon a spirit that lives in a lake. "How do you incorporate that value into a decision?" the water manager asked.[2] The ability of water managers to approach community values seems dependent upon whether that worth can be represented in a paradigm that is grounded in commodification.

Spiritual values of water

As it relates to spiritual or organic value, utility of water is highly intangible. From a community and spiritual standpoint, the value of water is difficult to quantify. Clearly resources can have very high value while having no monetary price. Market measurements of this resource actually limit understanding of its organic value, which is ultimately associated with sacred places in the environment.[3]

We found it an overwhelming task to identify how many people in the world actually hold non-monetary value for water. Our main concern was how to link religion or spirituality with the number of people who value water. First, we identified the religions and spiritual practices that value water and in what ways they value it. After that, we identified how many people worship in that religion or spiritual practice.

Many worldwide estimates tend to over- or under-represent adherents when those estimates are based in observation or an assumption that a person's geographic location determines religious beliefs. Because it relies on self-identification as its basis for classification, we found "Adherents.com" to be the most reliable source for establishing a relationship between spiritual beliefs and location. In addition to self-identification, the source lists variations from different sources for the number of adherents of particular religions worldwide, and it provides possible reasons for those differences. The creators of the site consider those and construct what they believe is a reasonable number of adherents. We use that number in this study.

Of approximately 7 billion people on the planet, about 84 percent consider themselves adherents to a particular religion. In the discussion below, we link the numbers of adherents to each of these religions that hold value in non-monetary ways for water. To better understand the importance of the non-market value of water worldwide, we discuss below the significance of water as it relates to spirituality, highlighting the meaning that various cultures assign to water and rites for which this natural resource is used. Through this discussion, it is apparent that a majority

DOI: 10.1057/9781137062499

of people worldwide place utmost importance on the organic or inherent value of water as it relates to spirituality. Accordingly, we argue that this is a value that should be incorporated into water development planning.

We considered approaching these beliefs in various ways, such as listing the religions from oldest to newest and also from most to least adherents. After much deliberation, we decided to address these religions from Eastern to Western, in order of the time that they were created—with the exception of ecologically based spiritual beliefs, which we address toward the end of this chapter. In general, these are religious or spiritual practices that revere water in certain ways or use it in ritual practices. Of course, it is impossible to include here all communities that identify with water in these ways. This section is simply meant to elucidate how many people in the world hold non-monetary values for water.

The Eastern beliefs we include are as follows: Jainism, Hinduism, Chinese traditional religion (including Taoism), Buddhism, and Shinto. With regard to Western spiritual beliefs, we reflect on Judaism, Islam, Christianity, and other spiritual and more ecologically oriented thought as addressed by Zoroastrianism and the Bahá'í faith. We also discuss similar ecological-based spirituality as embraced by some Native American and African groups.

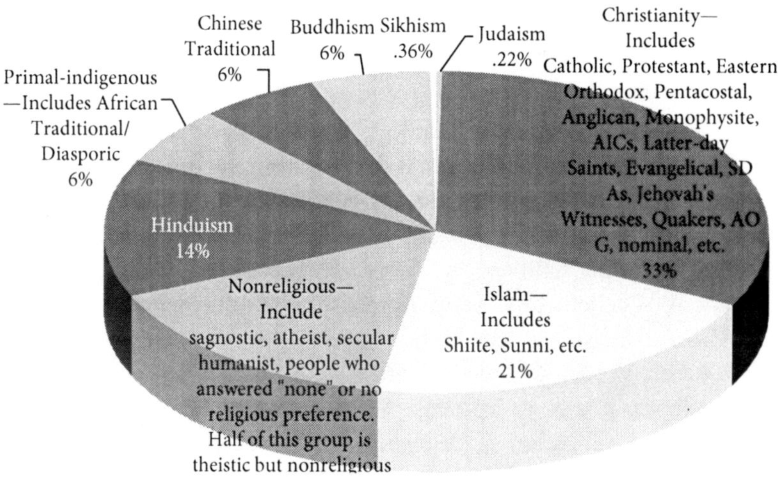

FIGURE 1 *Major religions of the world ranked by number of adherents*

Note: Due to rounding, total adds up to more than 100%.

Source: Adapted from Adherents.com.

DOI: 10.1057/9781137062499

> In most human cultures water has been variously used as a simile for the passage of time and change, or as a metaphor of higher realities such as the favor of the gods or the nurturing power of the sky... Everywhere it has been understood as a substance and a tool that can bring about greater and purer states of existence and remove what has soiled the quality of life.[4]

Eastern beliefs

Jainism

Jainism has approximately 4.2 million followers worldwide. It has a close relationship with Hinduism and Buddhism. Because all three of these religions have influenced each other in many aspects, Jainism has at times been considered a branch of both Hinduism and Buddhism. More recently, however, it is accepted to be a distinct religion of India. Moreover, it is "accepted that Jainism is an ancient religion of India and that it is older not only than Buddhism but also older than Vedic religion of the Hindus."[5]

> Water (*jala*) is one of the eight daily worship idols used as offerings, or *puja*. Water symbolizes the ocean. Every living being continuously travels through the Ocean of birth, life, death, and misery. This *puja* reminds that one should live the life with honesty, truthfulness, love and compassion towards all living beings. This way one will be able to cross the Life Ocean and attain *Moksha* or liberation.[6]

Hinduism

Even the origin of the name Hindu is water-related. The word comes from the Sanskrit word *Sindhu* and is derived from the Indus River, which originates in Tibet.[7] Hinduism crosses over different cultures. Although Hindu beliefs are typically associated with Indian culture, many communities, including Balinese and Bangladeshi societies, widely embrace this system of beliefs, too.[8] Large numbers of Hindus are also found in East Africa, Britain, and the United States. Approximately 900 million people adhere to Hinduism worldwide.[9]

This religion places an immense value on water. Hindus abide by a morning ritual obligation of cleansing. Fundamental in this practice is the ritual of taking a bath before entering a temple. The Ganges River is also central to many Hindus. Followers believe that bathing in the Ganges will wash away a person's impurities. It is for this reason that the Ganges is also referred to as God's divine vehicle and the savior of this world.[10] In addition, "Hindus... cast the ashes of their dead in the

DOI: 10.1057/9781137062499

river in the belief that this will guide the souls of the deceased straight to paradise."[11]

Chinese Traditional Religion

Approximately 394 million people follow Chinese traditional religion. Sometimes called Chinese folk religions, Chinese traditional religions are often categorized in different ways. Some scholars group Confucianism, Chinese Buddhism, and Taoism under the same label, while others identify them separately. There are convincing arguments for both approaches. "Even today there are very valid reasons for distinguishing Taoism from Confucianism, and distinguishing both from Chinese Buddhism and non-scriptural Chinese folk religion."[12] In this study, we recognize the distinctions made by Adherents.com, and we discuss these below.

Until the Cultural Revolution in China, Taoism had been one of the world's largest religions. At that time, however, the government moved to obliterate non-Communist religion by removing Taoist temples and clergy. The action was successful. Even though Islam, Buddhism, and Christianity have recuperated their losses, Taoism has done so less successfully. Accordingly, Adherents.com treats Taoism as a branch of Chinese traditional religions, and it looks at Buddhism separately. Confucianism adopts important ethical principles, but specifically does not prescribe to specific rituals or practices; therefore, we do not address Confucianism here.[13]

Because much of China is covered with mountains, the majority of people there are crowded into the river valleys, coastal zones, and deltas. Accordingly, many aspects of Chinese culture reveal value for water. Chinese Cosmology refers to theories concerning, and fundamental principles of, the cosmos and time-space relationships, and it closely reflects that country's religious and political changes. Stories of cosmology are related in historical ways, many of which include understandings of how Chinese culture evolved. They also involve human lessons.

One such tale involves a battle between flooding waters and the sky. The story closely parallels the flooding of the Yellow River, even today. In addition, the story describes water as primal and enigmatic; it needs to be controlled for the sake of the living.[14] The Xia, Shang, Zhou and Han Dynasty cosmologies all applied significance to water. The Han system stated that the universe operates in two forms, *yin* and *yang*. These

DOI: 10.1057/9781137062499

two modes interact, and they produce five types of basic energy: fire, water, earth, wood and metal. These arrangements of energy combine and recombine to produce the "ten thousand things" that constitute all phenomenal reality both perceivable and unperceivable.[15] The *Yin-Yang* symbolizes these dichotomous, yet interrelated, phenomena. Geomancy, or divination by means of geographical features, is also significant in Chinese culture. *Feng Shui*, the Chinese art of wind and water, became used to locate energy-filled sites for markets, cities, temples, shrines, and burial sites.

Practiced mainly in China, Tao translates to English as "The Flow of Things," the "Course of Nature," "The Way," or "The Watercourse Way."[16] Water is used as a representation of Tao because water always seeks the path of least resistance. It does not compete; it spreads out to find the easiest path, yet there is nothing stronger than water. It will carve through rocks, steel, and anything that resists it if there is no other path around, under or over it.

Buddhism

Like Hinduism, Buddhism crosses many cultural lines. Buddhism seems to have originated in the sixth century B.C. in northeast India. The religion has nearly 376 million followers worldwide, most of whom are in Asia. Buddhists do not tend to embrace symbolism and ritual, however, water is used in the funeral ceremony.[17] There are many variations of Buddhism currently found in many parts of the world, yet the tradition involving water is similar. Water fills a bowl and pours over the edge. As this occurs, the monks speak, "As the rains fill the rivers and overflow into the ocean, so likewise may what is given here [life] reach the departed."[18]

Shinto

Early Japanese practice was not called Shinto until Buddhism was established in the area. Shinto is considered Japan's indigenous religion, even though it exists more realistically as a historical and cultural set of practices. These practices are based on the veneration of the *kami*, the deities believed to inhabit natural phenomena such as mountains, trees, rocks, and springs. Whether it is public or private, worship of *kami* always begins with purification with water. Troughs for ritual washing are placed inside the many sacred shrines. In Shinto belief, waterfalls are believed to be sacred. Standing under them is believed to provide

DOI: 10.1057/9781137062499

purification.[19] There are approximately 4 million people worldwide who follow the practices set forth in Shinto.[20]

Sikhism

Sikhism was founded in the fifteenth century in the Punjab region of India. About 80 percent of Sikhs still live there. Sikhism used to be referred to as a break-off of Hinduism, but it is now seen as its own entity. There are approximately 23 million Sikhs in the world. In Sikh ritual water is used for baptism and the drinking of *Amrit*, or holy water that is prepared by stirring sugar and water with a double-edged sword.[21] *Amrit* is a Sanskrit word meaning "without death" or "immortal" and it is used in many types of ceremonies.[22]

Western beliefs

Judaism

Like Eastern faiths, Western beliefs also exhibit a wide array for the inherent worth of water and its consequent use for ceremonial worship. Below we address several of these faiths, and we show that across these belief systems there are similarities in organic values for water. Like other religions, Judaism crosses national borders. In general, its followers exist in clusters in different parts of the world. Although it is considered Israel's official religion, it has nearly 14 million adherents worldwide.[23]

Value for water is obvious in this religion. Like other religions, Judaism includes ritual washing. This washing can be as simple as washing hands and feet, or it can involve total immersion. The story of the Great Flood emphasizes punishment from the divine. In this story, the sole survivor, Noah, endures because of his moral values. The Red Sea is also prominent in Jewish history in that it allowed the Exodus of the Jews from the Egyptians. "The parting and crossing of the Red Sea shows that God has power over nature, even the mighty oceans. Water here is powerful, but is used as an instrument of God for punishment (for the Egyptians) and blessing (for the Israelites)."[24]

Islam

Between 610 and 632 A.D., the prophet Muhammad brought Islam to the polytheistic Arabs of Mecca.[25] Although the religion is predominant in Indonesia, Islam has nearly 1.5 billion followers worldwide.[26] In Islam,

DOI: 10.1057/9781137062499

water is important for purifying. Similar to the adherents of Shinto and Hindu faiths, Muslims must be ritually pure before they approach God in prayer. Some mosques have a courtyard with a pool of clear water in the center, and most mosques contain bathing areas outside the walls. Fountains, symbols of purity, are also sometimes found in mosques.[27] Islamic teachings emphasize the inherent right to water, underscoring its organic value.

> In the seventh century, the laws and codes of conduct of the then fledgling Muslim religion unambiguously articulated that access to fresh water is a right of all living beings... The idea that any person or group could control the availability of water or decide who gets water and who goes without was anathema and reprehensible to early Islam.[28]

Many Muslims today accept that recycled water can be used for agricultural and industrial purposes. In addition, it can be used for cleansing purposes "as long as its taste, color and smell have not changed."[29] With regard to prayer, the purity of water and its condition for use has been a matter of concern for some Muslims. This concern manifests itself in a resistance to the use of recycled water.

Christianity

Also practiced widely, the Christian faith has approximately 2.1 billion followers worldwide.[30] Use of water in ceremonial rites and for worship has changed over time. Because early Christians in Europe believed the use of water was a pagan practice, they banned the spiritual worship of water there. Saxon King Edgar even forbade the use of fountains in 960 A.D.[31]

Despite the ban on water worship, people's deep faith in the sacredness of water persisted... old customs were absorbed into Christian rituals and water worship hid behind a Christian façade. Water maintained its sacredness in rituals of baptism and hand washing.[32]

Today most Christian churches or sects have an initiation ritual involving the use of water. The symbolic ritual of baptism has its origins in the story of the Israelites being led by Moses through the Red Sea out of slavery in Egypt.[33]

Symbolism also emanates from the baptism of Jesus by John the Baptist in the Jordan River. According to Christian beliefs, Jesus commanded that his disciples baptize others after his resurrection. Different denominations of Christianity regard baptism in various ways. Except

DOI: 10.1057/9781137062499

for Catholicism, all Christian denominations conform to the idea that baptism does not in itself cleanse one from sin, but it is a declaration of a person's faith in Christ the Savior. In fact, Jesus described Himself as the "living water."[34]

In particular, the Armenian Church celebrates the Blessing of the Water on the day of Theophany, January 6. Called Churorhnek, the service is a commemoration of the baptism of Christ. The use of water in Christian symbolism is important in two ways. First, it indicates that just as they need water, Christians need God. Additionally, it symbolizes that everything is immersed in Him, a symbolic message of cleansing.[35]

Ecologically based spiritual beliefs

Zoroastrianism

Dating back to 1738 B.C., Zoroastrianism is an ancient religion of Iran. Although there are adherents to this religion worldwide, most of its 2.6 million followers exist in Iran and India.[36] Combining the properties of purification with its value as a primary element of life, Zoroastrianism embraces water as fundamental.[37] An intense belief in dualism and the opposition of good and evil guide the beliefs that pure water is sacred and polluted water is evil.

According to adherents of Zoroastrianism, Haurvatat, who is worshiped intensely, is the creator of water. Her being means wholeness, health, and integrity. During the holy day of Haurvatat in mid-summer, people pray and make offerings by the seashore or any natural water. "In everyday life Haurvatat is observed by keeping water unpolluted and being temperate and self-disciplined. Haurvatat is the personification of what salvation means to the individual."[38] Washing is also part of Zoroastrian ritual, and believers maintain a flood story similar to other religions.[39]

Bahá'í Faith

Founded in 1863 A.D. in Iraq, the Bahá'í faith revolves around the teachings of a Persian named Bahú'u'llá. The faith, which originates from Islam, includes laws concerning water and cleanliness. Specifically, it places great importance on the presence of water and its connection to agriculture and to ecological balance.

The Bahá'í administrative center is at Haifa, Israel, but the movement is widely spread in Europe, America, Africa and in Eastern countries. "For Bahá'ís, respect for the creation in all its beauty and diversity

DOI: 10.1057/9781137062499

is important, and water is a key element of that creation."[40] There are approximately seven million followers of this faith worldwide.[41]

Native American beliefs

Approximately 300 million people in the world follow an indigenous belief system. Adherents.com terms this category "Primal-indigenous." In order to challenge the connotation of the website's categorization as "primitive" or outside the realm of the norm, we identify this group simply as "Indigenous." Adherents.com includes "tribal religionists," "ethnic religionists," and "animists" within its category.[42] Also included in this group is "African traditional religionists," which is included under another subheading below. In addressing indigenous beliefs, it is important that one not make the assumption that these are exactly alike. Even though these beliefs often embrace environment as sacred, it is also important not to make the assumption that indigenous beliefs are necessarily congruent with the principles of environmentalism as these are broadly defined by the environmental movement.

"Many followers of Native American [s]pirituality do not regard their spiritual beliefs and practices as 'religion' in the way in which many Christians do. Their beliefs...form an integral and seamless part of their very being."[43] Nevertheless, some beliefs of indigenous groups in North America are similar to those of the Bahá'í faith in that there is an understanding of ecology and a respect for balance in life. About Native Americans and natural resource management, historian Norris Hundley, Jr. writes, "They recognized that, above all, they were part of nature and had to manage their own lives wisely."[44]

Tsalagi and Inuit belief systems provide examples of this balance. We also provide a glimpse at the ways in which indigenous beliefs about water can conflict with the dominant paradigm involving the discipline of economics and science. An issue involving the San Francisco Peaks allows a closer look at some beliefs of the Native tribes in Northern Arizona in the Southwestern United States.

Tsalagi

Tsalagi is the religion of the Amonsoquath, a member band of the Western Cherokee Nation. This belief system embraces a new day by "going to water." "Going to water" entails facing the rising sun and thanking Asgaya Galvlati for another new day and for the gift of a new sun.[45]

DOI: 10.1057/9781137062499

Inuit

The value of water for the Inuit is similar to the beliefs of other people in circumpolar regions such as Northern Russia and the Northern Scandinavian countries. Inuit religious belief is grounded in the idea that *anua* (souls) exist in all people and animals. In order for animals to continue to make themselves available to hunters, individuals, families, and the tribe must follow a complex system of taboos. There are many rituals and ceremonies that are performed before and after hunting expeditions to assure hunting success. Sedna or Takanaluk is the underwater goddess who rules the sea mammals. She is part human and part fish. She observes how closely the tribe obeys the taboos and releases her animals to the hunters accordingly.[46]

Native Beliefs in Arizona

In Northern Arizona, Native Americans have been involved in a controversy that they believe crosses lines of respect regarding their belief in the spiritual values. The issue stems from the use of reclaimed water by the Arizona Snowbowl, the local ski resort. Located on the San Francisco Peaks north of Flagstaff, the Snowbowl, which opened in 1938, was controversial from its inception; nevertheless, it has thrived and has existed as part of the local economy. More than 13 Native American tribes throughout the Southwest consider these mountains sacred.

Officials of the Forest Service have approved using recycled water for snowmaking on the Peaks.[47] This decision and the consequent court decisions upholding the decision are considered an outrage to the Native Americans in the area because they believe impure and unnatural water defiles their blessed place. This statement sums up Navajo tribal views: " 'To' bee iina"—"Water is life."[48]

African spirituality

Many of Africa's indigenous beliefs also carry with them the essence that God exists in the form of nature. Mbiti, an African scholar states,

> The great creator has very few temples or images, but is almost everywhere believed in... Ordinarily there is little direct contact between man and the Supreme deity... Below God are the spirits, which Mbiti describes as invisible, ubiquitous, and unpredictable. They live everywhere—in the sky, the sun, the earth, bodies of water, rocks, or trees."[49]

DOI: 10.1057/9781137062499

Additional ways people hold spiritual value for water

The list above is obviously not comprehensive or conclusive; however, it provides concrete examples of the number of adherents of religious or spiritual beliefs worldwide that hold value for water. For obvious reasons, knowing when to stop can be as challenging as knowing where to start. We have chosen to stop this list at adherents of religions that number a million followers or fewer, even though some of these, such as Rastafarianism and some Neo-pagan beliefs such as Wicca spirituality hold obvious uses for water.[50]

Because religious and spiritual beliefs about water cross so many cultural lines and, therefore, might be counted twice, we have also chosen not to address the huge number of specific sites where people worship or hold spiritual value for water. These could include vast numbers of holy wells or springs such as those in Ireland, England, Turkey, or Tibet. This list would also include a considerable number of rivers and lakes, such as the Ganges and Yamuna Rivers, both in India, Lake Titicaca in Bolivia, Lake Bosumtwi in Ghana, and Grand Lake in Colorado in the Southwestern United States.

These examples briefly touch upon the number of people in the world who hold spiritual beliefs about water, yet through this discussion, we begin to grasp some intangible values for water that people worldwide hold. In this respect, market systems rarely reflect the true utility of water. Through awareness and appreciation of these inherent values, water managers can consider the social and cultural consequences of their decisions. Below, we discuss reasons water managers might want to do so, and we explore ways in which policy makers can deem community values as fundamental.

Notes

1 Tom Holm, Diane J. Pearson, and Ben Chavis, "Peoplehood: A Model for the Extension of Sovereignty in American Indian Studies," *University of Minnesota Press, Wicazo Sa Review*, Vol. 18, Issue 1, (Spring 2003), pp. 7–24, 14. http://www.jstor.org/stable/1409431.

2 Kira Artemis Russo, Personal Interview, Canada, October 16, 2012.

3 Holm *et al.*, 14.

DOI: 10.1057/9781137062499

4 J. Goulde, "Water in Classical Chinese Religion," *Water: The Mystery, Art and Science of Water.* Sweet Briar College, (2004), accessed November 30, 2004, http://witcombe.sbc.edu/water/introduction.html.

5 "Antiquity of Jainism," Jainworld.com (2011), accessed December 20, 2012. http://www.jainworld.com/jainbooks/antiquity/jainorel.htm.

6 Hansa Sutaria and Vinod Sutaria, "Jain Rituals and Ceremonies," accessed December 20, 2012. http://www.fas.harvard.edu/~pluralsm/affiliates/jainism/workshop/Sutaria%20Jain%20Rituals.pdf.

7 Satguru Sivaya Subramuniyaswami, *Dancing with Siva: Hinduism's Contemporary Catechism* (Himalayan Academy Publications, 2003), 1008.

8 "Balinese Hinduism: A Life of Ritual and Devotion," Indonesia-Bali, 2004, accessed November 30, 2004. http://www.indonesia-bali.com/bali_hindu.htm.

9 "Hinduism," Adherents.com National and World Religion Statistics (2007), accessed December 19, 2012. http://www.adherents.com/Religions_By_Adherents.html#Hinduism.

10 M. Guin, "Ganga: River and Goddess," *Dolls of India* (2004), accessed November 30, 2004. http://www.dollsofindia.com/ganga.htm.

11 Abrams.

12 "Chinese Traditional Religion," Adherents.com: National and World Religion Statistics, 2007, accessed December 19, 2012. http://www.adherents.com/Religions_By_Adherents.html#Chinese.

13 "Confucianism," ReligionFacts (2012), accessed December 19, 2012. http://www.religionfacts.com/a-z-religion-index/confucianism.htm.

14 Goulde.

15 Ibid.

16 "Taoism/Daoism," *Yakrider* (2003), accessed November 30, 2004. http://www.yakrider.com/Tao/Taoism_Daoism.htm.

17 V. Aubuchon, *Vaughn's One-Page Summaries*, Vaughn's One-Page Summaries, accessed January 25, 2005 http://www.vaughns-1-pagers.com/#religion; I. Hexham. *Concise Dictionary of Religion.* Regent College Publishing, University of Calgary (2000), accessed January 25, 2005. http://www.acs.ucalgary.ca/~nurelweb/books/concise/; W. Swatos, Jr, *Encyclopedia of Religion and Society*, Hartford Institute for Religion Research (2005), accessed January 25, 2005. http://hirr.hartsem.edu/ency/index.html#B.

18 Abrams.

19 J. Bowker (ed.), *The Oxford Dictionary of World Religions* (Oxford: Oxford University Press, 1997), http://www.africanwater.org/religion.htm..

20 "Shinto," Adherents.com: National and World Religion Statistics (2007), accessed December 19, 2012. http://www.adherents.com/Religions_By_Adherents.html#Shinto.

DOI: 10.1057/9781137062499

21 "Amrit Ceremony," Religions, BBC, October 27, 2009, accessed December 20, 2012. http://www.bbc.co.uk/religion/religions/sikhism/ritesrituals/amrit.shtml.

22 "Amrita," Wikipedia, December 10, 2012, accessed December 20, 2012. http://en.wikipedia.org/wiki/Amrita.

23 "Judaism," Adherents.com: National and World Religion Statistics (2007), accessed December 21, 2012. http://www.adherents.com/Religions_By_Adherents.html#Judaism.

24 Bowker.

25 Aubuchon.

26 "Islam," Adherents.com: National and World Religion Statistics (2007), accessed December 21, 2012. http://www.adherents.com/Religions_By_Adherents.html#Islam.

27 J. Rothfeder, *Every Drop for Sale: Our Desperate Battle Over Water in a World about to Run Out* (New York: Penguin Putnam, Inc., 2001).

28 Ibid.

29 "Pakistan Link," "Issues on Recycled Water and Copyrights," Pakistan Link (2004), accessed November 30, 2004. http://www.pakistanlink.com/religion/99/06-04.html.

30 "Christianity," Adherents.com: National and World Religion Statistics (2007), accessed December 21, 2012. http://www.adherents.com/Religions_By_Adherents.html#Christianity.

31 Vandana Shiva, *Water Wars: Privatization, Pollution, and Profit* (Cambridge: South End Press, 2002), 136.

32 Ibid., 137.

33 Abrams.

34 *Holy Bible, Revised Standard Version*,. Division of Christian Education of the National Council of the Churches of Christ in North America, ed. (Camden, NJ: Thomas Nelson, 1946), John 4:1–42.

35 "Ceremony of the Blessing of the Water—Churorhnek," *Armenian Heritage* (2004), accessed November 30, 2004. http://www.armenianheritage.com/rewater.htm.

36 Swatos, Jr.; "Zoroastrianism." Adherents.com: National and World Religion Statistics (2007), accessed December 21, 2012. http://www.adherents.com/Religions_By_Adherents.html#Zoroastrianism.

37 Hexham.

38 Bowker.

39 "Flood Myths," *Morgana's Observatory* (2004), accessed January 25, 2005. http://www.dreamscape.com/morgana/titania.htm.

40 Arthur Lyon Dahl, "The Bahá'í Perspective on Water." *International Environment Forum* (April 18, 1999), accessed January 25, 2005. http://www.bcca.org/ief/ddahl97e.htm.

DOI: 10.1057/9781137062499

41 "Bahá'í," Adherents.com: National and World Religion Statistics (2007), accessed December 21, 2012. http://www.adherents.com/Religions_By_Adherents.html#Baha'i.
42 "Primal-indigenous," Adherents.com: National and World Religion Statistics (2007), accessed December 21, 2012. http://www.adherents.com/Religions_By_Adherents.html#primal.
43 B.A. Robinson, "Native American Spirituality," *Ontario Consultants on Religious Tolerance*, (December 8, 2004), accessed November 30, 2004. http://www.religioustolerance.org/nataspir.htm.
44 Norris Hundley, Jr., *The Great Thirst: Californians and Water, 1770s-1990s* (Berkeley: University of California Press, 1992), 15.
45 "Tsalagi Religion," Sovereign Amonsoquath Tribe of Cherokee (1994), accessed November 30, 2004. http://amonsoquath.addr.com/tsalagi_religion.html.
46 Robinson.
47 "Draft Environmental Impact Statement for Arizona Snowbowl Facilities Improvement," *Coconino National Forest*, accessed December 24, 2004. www.fs.fed.us/r3/coconino/publications/snowbowl/.
48 John Bianchini, "'To' bee iina" Water is life," *Navajo-Hopi Observer* (2004), accessed November 30, 2004. www.navajohopiobserver.com/news1.htm.
49 "African Religion," *The Drum* (Spring 1995), accessed December 24, 2004. http://www.ritesofpassage.org/ds95-2.htm.
50 "Rastafari Movement," *Wikipedia* (December 20, 2012), accessed December 27, 2012. http://webcache.googleusercontent.com/search?q=cache:Cu4QTVfL3g4J:en.wikipedia.org/wiki/Rastafari_movement+rastafarian+symbols+water&cd=7&hl=en&ct=clnk&gl=us; "Wicca and Witchcraft," ReligionFacts (December 21, 2012), accessed December 27, 2012. http://www.religionfacts.com/neopaganism/paths/wicca.htm.

DOI: 10.1057/9781137062499

5

Cooperative Communities: The Future of Water Management

Abstract: *This chapter focuses on the human right to water. Over the past 50 years, the UN has defined human rights, many of which concern issues of environmental justice. UN General Comment No. 15 served to refine the human right to water, recognizing it as "a public good fundamental for life and health." It is estimated that women spend a combined total of 200 million hours per day collecting water, making them subject to violence on long treks and susceptible to water-borne illness. Other global initiatives have addressed issues of gender equality and equality for minorities. We summarize our suggestions for water managers. Representatives of public trust, water managers ought to represent community values. In doing so, they can serve the tenets of "water democracy."*

Keywords: Human right to water, General Comment No. 15 and water, gender equality and water, water management, water managers as agents of public trust, "water democracy"

Russo, Kira Artemis and Smith, Zachary A. *What Water Is Worth: Overlooked Non-Economic Value in Water Resources.* New York: Palgrave Macmillan, 2013.
DOI: 10.1057/9781137062499.

Assignments of non-monetary value

As we stated in Chapter 2, commodification of water began with the introduction of the Doctrine of Prior Appropriation, which provided absolute water rights to ownership of property. Within this value system lay the capitalist view that all values can be exchanged. Moreover, commodification of water has become the dominant paradigm. Prior appropriation lends to market competition, and supporting government institutions maintain the status quo. Those with the strongest financial influence are, thereby, allowed to prevail.

A social construct based mainly on property ownership, commodification shapes ways in which values of water are taken into account. In other words, the value system is manipulated through commodification. Ultimately, this paradigm imparts the idea that values that can be quantified should be included. When values cannot be easily quantified (such as certain community values), they often cannot be included. Non-inclusion means that cultural beliefs might be overlooked, suppressed, or even subjugated.

As we began our research, we set out to find whether the acceptance of this paradigm by water managers might unfold such that, even if they *want* to include non-monetary values, the legal and social frameworks under which they work might not allow for this inclusion. Non-monetary values are those not specifically related to production.

Russo's survey revealed four types of non-monetary values. These are as follows: environmental uses, which include uses for nature and animals, conservation, bathing, and recreation. As one would expect, some of these values are vague. For example, environmental uses were described as both monetary and non-monetary values. They leaned toward monetary uses such as tourism or fish-farming, and they were valued as such. Recreation also conformed to the standards associated with tourism, and use of water for bathing can be considered an organic value or an indication of water quality for domestic use. Without a prompt, all of the water managers stated that they use water in ways that reflect monetary values. These uses are prominent over non-monetary uses in all the countries represented in that survey. In no case or response were non-monetary uses stated first. Even though some water managers referred to environmental values, the underlying assumption then is that they are not priority uses.

DOI: 10.1057/9781137062499

This study also revealed that conservational values for both water quantity and quality were important to water managers. There is generally high utility for water for future generations. As stated in Chapter 1, discount rates allow this value to be commodified. In other words, cultural conservation can be protected through the status quo involving commodification. What is less clear, however, is how strongly community ethics of conservation weigh against other values. Even though conservational values for both quantity and quality are deemed important, there is little indication that water managers actually address this value. As stated above, none of the respondents disregarded monetary values.

Assignments of non-monetary values and emergent ecological benefits

Foster asserts that exploitation of the environment has become more universalized because the elements of nature are increasingly subjected to measures of profitability.[1] Ecologically, commodification of water lends to water managers making the general assumption that water has the same value in each local ecological system. To the contrary, people hold different values for water even within the same bodies and along certain stretches of it. "The generic value of hydrologic services is apparent, but the functionality and value of an ecosystem is likely to be highly variable, so site-specific assessment remains important."[2] Brauman believes the solution toward ecological benefits lies in the integration of valuation methods, which are based on the cost of both production and social values. "These integrative values are more likely to capture the full value of ecosystems in providing services."[3]

Ecological benefits also lie within interdisciplinary and ecologically based approaches to science. Adrian Muller argues for post-normal science, science that attempts to be aware of normative assumptions. This would also encompass trans-disciplinary values.[4] Richard B. Norgaard combines the call for interdisciplinary work with method. He believes that this crossover involving methodology allows for methodological pluralism.[5] The ability to apply ecological benefits hinges on both quantitative and qualitative measurements.

At the core of change in both the financial and scientific realms is institutional change and an understanding from policy makers, including water managers, that ecological benefits are coupled with social justice.

DOI: 10.1057/9781137062499

A.Y. Pagdee *et al.* state that markets cannot measure equity.[6] We look further at equity and social benefits below.

Implementation

Financial views and instrumentalism are incorporated into almost all policy decisions in the world today; therefore, it is difficult to dissociate from this outlook in order to implement water policy that reflects non-monetary values. In his book *The Environmental Endgame: Mainstream Economics, Ecological Disaster, and Human Survival*, Robert L. Nadeau argues that there is no basis for assuming that the discipline of economics is scientific, nor that it is a rigorously mathematical discipline comparable to physics. Rather, he contends that market economics is based upon metaphysical assumptions. "The creators of classical economic theory ... firmly believed that the natural laws of economics originated in the perfect mind of the Creator of a mechanistic universe."[7] Robert Paelke elaborates, "We must discover ways to leave in place the real efficiencies of markets while inserting nonmarket values such as intergenerational equity and sustainability."[8]

The ability to implement comprehensive water policy is further obstructed by erroneous beliefs about markets. In separate works, Nicholas Georgescu-Roegen and Herman Daly and John B. Cobb identify certain fallacies embedded within the discipline of market economics. Daly and Cobb link market economics and environment to Alfred North Whitehead's fallacy of misplaced concreteness: Conclusions about the real world are drawn from abstractions, and dangers are not fully examined.[9] Georgescu-Roegen recognizes the fallacy of endless substitution. Market economics thrives on the notion that commodities can be replaced by another when they are depleted, yet in the case of water, there is no substitution. Poverty arises from this sort of exploitive water marketing.[10]

Foster contends that the "successes" of the market have led from regional to global destruction.[11] Market competition means that the "economically powerful" can invest in capital-intensive means to appropriate water regardless of community values and the limits of water systems. Further, water marketing encourages private interest groups to ignore community control over water.[12] Institutions associated with global trade encourage neoliberal views that promote market economics and less government involvement, and the bureaucratic status of water

DOI: 10.1057/9781137062499

managers justifies the status quo of class domination, regardless of the manager's intent. The obstructions to implementation of water policy that includes community values are obvious.

While privatization is generally couched in rhetoric about the disappearing role of the state, what we actually see is increased state intervention in water policy subverting community control over water resources. Policies imposed by the World Bank, and trade liberalization rules crafted by the World Trade Organization (WTO) , are creating a sweeping culture of corporate-states all over the world.[13]

When monetary motivations become political, community values are often plainly disregarded. An example of this preference is hydraulic fracturing, or fracking. In these cases, policy allows for oil companies to fracture the geologic base of aquifers in search of oil. This fracturing often causes methane contamination of water resources and even radioactive contamination of them. Some people in the areas where fracking occurs can literally light their tap water on fire. This instance provides an interesting look at economic exchange values for natural gas versus the use values of water. With regard to implementation of policies that include non-monetary values of water, the paradox is obvious.

Globalization has standardized many things on the planet. People all over the world now share values and ideas that not long ago were culturally specific. Some of these changes have been good—others less so. The dominance of markets in water management, particularly in semi-periphery and periphery countries, is a value that might not be worth importing. We argue that even the best monetary instruments should only be one of the tools competent water managers use in their decision making.

Water as a human right

In Chapter 3 we identified the number of people in the world who are affected by limited water supplies. To reiterate, water scarcity affects approximately one-fifth of the global population. The UN estimates that about 1.2 billion people have limited water supplies. Another half billion live in areas approaching scarcity. An additional 1.6 billion people lack the infrastructure that will allow them to extract water from rivers and aquifers. UN Secretary-General Ban Ki-moon states, "All [locations of

DOI: 10.1057/9781137062499

scarcity] are places where shortages of water contribute to poverty."[14] He asserts that water shortages create tension in these regions.

Over the past 50 years, the UN has further defined human rights, many of which have been concerned with issues of environmental justice. Nevertheless, it was not until November 2002 that the UN Committee on Economic, Social and Cultural Rights issued a report called "Substantive Issues Arising in the Implementation of the International Covenant on Economic, Social and Cultural Rights," better known as General Comment No. 15, which served to refine the human right to water. In the first paragraph of the document, the committee recognizes water as "a public good fundamental for life and health."[15] Further, it states, "The human right to water is indispensible [sic] for leading a life in human dignity. It is a prerequisite for the realization of other human rights."[16]

Overall, General Comment No. 15 protects for both personal and domestic use the following rights of people worldwide: sufficient, safe, acceptable, physically accessible, and affordable water. According to the UN, sufficient supplies of water equal the amount that would ensure that most basic needs are met and that few health concerns arise. The UN declares this amount to be between 50 and 100 litres of water per person per day.[17] The UN defers to the World Health Organization guidelines for safe drinking water quality. The term "acceptable" embodies the idea that water is used in culturally appropriate ways and that it is suitable and sensitive to the needs of gender, lifecycle, and privacy. In addition, the UN states that water should be within 1,000 meters of the home, that collection time should not exceed 30 minutes and that water costs should not exceed three percent of household income.[18]

In September 2010, the UN Human Rights Council created the most far-reaching resolution to date regarding the human right to water. It stated that international resolutions created by the UN regarding these rights are legally binding upon countries worldwide.[19] In other words, states are responsible for carrying out international law regarding the human right to water. The resolution asks that states develop "appropriate tools and mechanisms to achieve progressively the full realization of human rights obligations related to access to safe drinking water and sanitation, including in currently unserved and underserved areas."[20] This delegation of power from the Human Rights Council characterizes the challenges associated with addressing global problems and practical

DOI: 10.1057/9781137062499

solutions, especially among nations where monetary resources to develop water supplies are scarce.

It is evident that questions involving the human right to water will not easily be solved. In a world where water supplies are dwindling, competition for water resources is also increasing, and the discussions regarding these resources are becoming more politically charged. As such, it is obvious that deliberation continue at both the global and state levels with regard to this topic. In the next section, we look at the relationship between gender, the inclusion of community values of water into policy and human rights.

The role of gender

According to Women's Earth Alliance, "Women worldwide spend a combined total of 200 million hours per day collecting water."[21] In some areas, women and children spend up to eight hours per day collecting water. Not only are they unable to do other paid work or attend school, but women are often subject to violence on these long treks to obtain water. If water supplies are unavailable or contaminated, they might spend even more time and labor collecting this resource. In addition, water-related diseases are common for women because they have close contact with the water and because they often are responsible for caring for the sick.[22]

General Comment No. 15 states, "Whereas the right to water applies to everyone, States parties should give special attention to those individuals and groups who have traditionally faced difficulties in exercising this right, including women, children, minority groups, indigenous peoples, [and] refugees."[23] It adds that particular steps should be taken to see that women are not excluded from decision making processes about water and entitlements. "The disproportionate burden women bear in the collection of water should be alleviated."[24]

Much attention has been given to an increased role in water management by women, yet this glaring disparity in human rights is still obvious. This problem was first recognized at the global level in 1977 at the UN-Water Conference at Mar del Plata. In January of 1992, the International Conference on Water and the Environment in Dublin explicitly recognized the fundamental role that women play in the provision and safeguarding of water. This conference also made reference for the need for women in the management of water resources. Agenda 21

DOI: 10.1057/9781137062499

and the Johannesburg Plan of Implementation support this global reference to women in this field. "Without specific attention to gender issues and initiatives, projects can reinforce inequalities between women and men and even increase gender disparities."[25]

Frances Cleaver and Diane Elson suggest that, in reality, international policy statements do little to enhance women's powers. They suggest that local water policy should be "gender aware."[26] In addition, they suggest institutional analysis of both formal and informal institutions. Among other human rights, they argue that institutional change should address women's time as an economic good. They believe this would make obvious the current undervaluation of women's time. Finally, they suggest that women organize for change.[27]

Education worldwide

Bruce Yandle, Maya Vijayaraghavan, and Madhusudan Bhattarai assert that rising incomes make it possible for communities to build institutions that create and protect advanced property rights.[28] With regard to water policy, the paradoxical nature of their argument is evident: The creation of private property fosters commodification of this resource and lends to the ability to trade in a market system. However, the importance of the arguments of Yandle *et al.* lies mainly in ideas involving equality and social justice. Education is essential for disentangling issues involving the human right to water, gender equality, and justice.

> Improving the environment is not free; opportunity costs matter to real people. How it happens is a story about growing incomes and environmental use. But it is not just a story about income and the environment. Income increases alone will not bring changes in environmental quality. Rising incomes become the means for making institutional changes that will conserve and, in some cases, rebuild environmental quality. Income-driven institutional change is costly, but not as costly as an unbounded tragedy of the commons.[29]

Mariano Torras and James K. Boyce suggest that the crown jewels of institutional change involve "vigilance and advocacy."[30] They argue that in the absence of these, greater output can mean greater consumption or waste. "Why might 'vigilance and advocacy' ... in a society increase with per capita income? One possibility is that individual demand for environmental quality rises. Another is that individuals gain greater power to make that demand effective through the political process."[31]

DOI: 10.1057/9781137062499

Water managers as agents of public trust

Our arguments about human rights, gender equality, and social justice have made clear the responsibility water managers hold. In Russo's study, water managers identified the following as impediments to incorporating the values outside monetary values that they want to include: lack of money, lack of community support, political interference, lack of transparency, and lack of accountability. Political interference weighed heavily in this realm; in fact, this plays a role in all the obstacles mentioned.

Considering both monetary and non-monetary values of water, we outline here five ways in which water managers could include community values in decisions about water policy. These approaches imply integration on all levels. There are likely others, but these suggestions provide a solid foundation for water managers. Some of these practices are already integrated into water management. We note below when these inclusions are nominal rather than substantive.

1 Some community values can be accommodated within conventional monetary methods. In this way, water management already conforms to the paradigm of commodification. As we stated above, community inclusion in institutions is key.
2 Water managers can set in motion discussions with communities about inherent or organic values. These discussions already exist in policies such as Environmental Assessments (EAs), however, these deliberations must be of substance in order to accommodate community values.
3 Related to the previous suggestion, water managers can include Traditional Knowledge, so that organic values can truly be represented. Again, this must be in deed and not simply in name.
4 Water managers could be selected in such a way that they represent community interests and are "gender aware." This might allow different views that traverse the dominant paradigm of commodification. In some cases, this might be through election. Appointment might also be an option.
5 There could be a selection of more than one water manager.

The fourth and fifth suggestions obviously could be politicized, but they address decision making that counteracts the forces of institutionalism, which often represents views prevalent within the dominant para-

DOI: 10.1057/9781137062499

digm. Additionally, these potential solutions lend to participation and deliberation.

Ultimately, these suggestions provide the necessary framework for what Gramsci called a war of position against Cultural Hegemony. He believed that breaking the cycle of consent was necessary for change. Because compliance lacks power, people could never engage in a passive revolution. Breaking the cycle of consent means that people no longer accept the institutions from which the "common sense" is devised and that education becomes a pervasive force to counter hegemonic influences.[32]

Conclusion

Our intent here has been to show how many people worldwide value water in non-monetary ways. We have also approached whether water managers are able take into account the values of their community. Our book allows researchers to look further into the ways in which the abilities of water managers to represent society might be guided or limited by the financial practices of hegemonic powers. Understanding the difference between social values and managerial practice allows us a new place to begin the discussion regarding trust in government and justice as fairness in the field of water policy.

Water managers worldwide seem to be renouncing—in favor of market solutions—the values placed on non-monetary features of water management. We argue that monetary standards often fail to align government with community values; further, we assert that this lack of inclusion minimizes trust in government. This study is not simply about cultural sensitivity; it is meant to serve the interest of water managers. It deals with the implications of conforming to standards that address only monetary values. Trust in government extends from the recognition of cultural values and the human right to water. These are critical long-term goals. Shiva defines this integration as "water democracy."[33]

Here we are not implying that we completely remove market values from water management; that argument would be unrealistic. Rather, we argue that the concept of value and benefit to society should be fully served. This would mean that water policy and market interests are dealt with at a more local level where we can truly assess the benefits to

DOI: 10.1057/9781137062499

communities. Ultimately, this approach would allow us to best utilize our resources so that we can effectively deal with issues of waste and scarcity to ensure the protection of this vital resource.

In theory, price should measure use value; however, price is not necessarily tied to exchange value. In terms of water management, we contend that its political exchange value through commodification is, in actuality, higher than its use value. Waste and degradation are often the result of skewed views of value. We have laid the groundwork for this understanding, and we support further research on this topic.

We live in a world where monetary values seem to be applied to everything. ("Is going to college worth the investment?" "Will a major in business lead to more income than a major in history?") There is solid research that having and making money does not lead to a happier and more fulfilled life.[34] Such weighty concerns are beyond the scope of this book, but we think it is worth pointing out that globalization and economic and political forces are what left us, in a word, where markets and monetary values determine the worth and allocation of nearly everything. It might behoove water and other natural resource managers to step back and consider all the ramifications of their resource management decisions. The authors of this book live in Arizona, a state in the Southwestern part of the U.S. This state does not recognize in-stream use as a beneficial use of water. Much of the developing world is also adopting the beneficial use standard as the means by which they allocate water resources. In Arizona the rivers and streams are all fully appropriated—all for beneficial economic uses. There are some things in life that have no price.

> "What's water but the generated soul?"
>
> W.B. Yeats, "Coole Park and Ballylee," 1931

Notes

1 John Bellamy Foster, *The Vulnerable Planet A Short Economic History of the Environment*, (New York: Monthly Review Press, 1994).

2 Kate A. Braumann, Gretchen C. Daily, T. Ka'eo Duarte and Harold A. Mooney, "The Nature and Value of Ecosystem Services: An Overview Highlighting Hydrologic Services." *Annual Review of Environment and Resources*, Vol. 32, Issue 1 (2007): 159.

3 Ibid., 146–147.

DOI: 10.1057/9781137062499

4 A. Muller, "A Flower in Full Blossom? Ecological Economics at the Crossroads Between Normal and Post-normal Science," *Ecological Economics*, Vol. 45 (2003): 19–27.

5 R. B. Norgaard, "The Case for Methodological Pluralism," *Ecological Economics*, Vol. 1, Issue 1 (1989): 37–58.

6 A. Y. Pagdee, Y. S. Kim, and P. J. Daugherty, "What Makes Community Forest Management Successful: A Meta-study from Community Forestry Throughout the World," *Society and Natural Resources*, Vol. 19, Issue 1 (2006): 43.

7 Robert L. Nadeau, *The Environmental Endgame: Mainstream Economics, Ecological Disaster, and Human Survival* (New Brunswick, NJ: Rutgers University Press, 2006), xi.

8 Robert Paelke, *Environmentalism and the Future of Progressive Politics* (New Haven, CT: Yale University Press, 1989), 141.

9 Alfred North Whitehead. Science and the Modern World (NY: Free Press, 1997), 51; H.E. Daly and J. Cobb, *For the Common Good: Redirecting the Economy Toward the Community the Environment and a Sustainable Future* (Boston: Beacon Press, 1989), 35.

10 Vandana Shiva, *Water Wars: Privatization, Pollution, and Profit* (Cambridge: South End Press, 2002), 15.

11 Foster, 1994, 35.

12 Shiva, 23–24.

13 Ibid., 87.

14 Ban Ki-Moon, "Address as Prepared for Delivery to the Davos World Economic Forum," January 24, 2008, accessed September, 2011. http://www.un.org/apps/news/infocus/sgspeeches/search_full.asp?statID=177

15 Committee on Economic, Social and Cultural Rights, Twenty-ninth Session, "Substantive Issues Arising in the Implementation of the International Covenant on Economic, Social and Cultural Rights" (General Comment No. 15), Geneva, Switzerland, November 11–29, 2002, 1, accessed September, 2011. http://www.unhchr.ch/tbs/doc.nsf/0/a5458d1d1bbd713fc1256cc400389e94/$FILE/G0340229.pdf

16 Ibid.

17 Ban Ki-moon.

18 Ibid.

19 UN-Water Decade Programme on Advocacy and Communication (UNW-DPAC), "The Human Right to Water and Sanitation: Milestones," (2011), 3, accessed September, 2011. http://www.un.org/waterforlifedecade/pdf/human_right_to_water_and_sanitation_milestones.pdf

20 Ibid.

21 Women's Earth Alliance, "Water," accessed May 15, 2013. http://globalwomenswater.org/.

22 Ibid.

DOI: 10.1057/9781137062499

23 Committee on Economic, Social and Cultural Rights, Twenty-ninth Session, 7.

24 Ibid.

25 United Nations Department of Economic and Social Affairs (UNDESA), “International Decade for Action ‘Water for Life’ 2005–2015,” accessed May 15, 2013. http://www.un.org/waterforlifedecade/gender.shtml.

26 Frances Cleaver and Diane Elson, “Women and Water Resources: Continued Marginalisation and New Policies,” Sustainable Agriculture Programme of the International Institute for Environment and Development, (1995), 1, accessed May 15, 2013. http://www.mekonginfo.org/assets/midocs/0003073-environment-women-and-water-resources-continued-marginalisation-and-new-policies.pdf.

27 Ibid., 8.

28 Bruce Yandle, Maya Vijayaraghavan, and Madhusudan Bhattarai, “Income and the Race to the Top.” *You Have to Admit It's Getting Better* (Stanford, CA: Hoover Institution Press, 2004), 86.

29 Ibid., 87.

30 Mariano Torras and James K. Boyce, “Income, Inequality, and Pollution: A Reassessment of the Environmental Kuznets Curve,” *Ecological Economics,* 25 (1998), 150, accessed March, 2013. ftp://131.252.97.79/Transfer/ES_Pubs/ESVal/EnviroKuznetCurve/torras_kuznetandequality_1998_ecolecon_v25_p147.pdf.

31 Ibid.

32 Anne Showstack Sassoon, ed., *Approaches to Gramsci* (London: Writers and Readers Publishing Cooperative Society Ltd., 1982), 141.

33 Shiva, 27.

34 See e.g. David G. Myers, “The Funds, Friends, and Faith of Happy People,” *American Psychologist,* Vol. 55, Issue 1, (January 2000): 56–67; Aaron Ahuvia, “If Money Doesn't Make Us Happy, Why Do We Act As If It Does?” *Journal of Economic Psychology,* Vol. 29, Issue 4 (August 2008): 491–507; David G. Myers and Ed Diener, “Who Is Happy?” *Psychological Science,* Vol. 6, No. 1 (January 1995): 10–19.

DOI: 10.1057/9781137062499

Glossary of Terms

commensuration allows us to compare, as in cost-benefit analysis, two different entities

commoditization the way in which goods become discernible among others in the marketplace

commodification the process by which value is assigned

"common sense" Aristotle's term for ideas that become the social norm

contingent valuation (CV) based on hypothetical scenarios, a method to assess non-use (non-market) values of natural resources

core nations Countries at the core of production. Core countries hold dominance as productivity dominance, trade dominance, and financial dominance.

cost-benefit analysis (CBA) a method of assigning dollar values to non-monetary values so that they can be traded

Cultural Hegemony A theory proposed by Antonio Gramsci based on Marxist theory. Gramsci argued that political and social preferences do not develop solely from economic struggle. These preferences also reflect assumptions about how society already exists and the normative aspect of how society should be.

demand-side economics A theory of the discipline of economics that attempts to address the inefficiencies associated with private sector decision making. Keynesian economists argue that one role of government is to manage the effects of these inefficiencies.

depreciation a decrease in value over time

Ecological Economics a sub-discipline of economics that is based, to a great extent, upon the notion that the conventions of a market economy fall short when it comes to understanding human behavior

economic Paul Samuelson's term describing how society ends up choosing, *with or without the use of money*, to employ scarce productive resources that could have alternative uses, now or in the future [emphasis added]. This term also denotes a distribution of wealth or an accumulation of goods or services.

economic good something that satisfies a need or want

ecosystem services the benefits people attain from ecosystems

elasticity a measure of a variable's sensitivity to change as it relates to another variable

Environmental Economics a sub-discipline of economics, this field applies financial theory to environmental problems

exchange-value/trade value Karl Marx's term that describes the idea that two commodities of unequal use-value (see below) can be reduced to a third non-commodified value that is different from the use-values of either of the commodities. In other words, two commodities of unequal value can be traded through the use of a third, substituted, value.

fallacy of misplaced concreteness Alfred North Whitehead's theory that conclusions about the real world are drawn from abstractions, and dangers are not fully examined

fallacy of endless substitution Market economics thrives on the notion that commodities can be replaced by another when they are depleted. In the case of water, there is no substitution.

hydrologic/hydrological services the benefits people attain from water

Institutional Theory highlights the ways in which associations guide human behavior

International Monetary Fund (IMF) conceived at the Bretton Woods Conference in 1944 to avoid a recurrence of the cycle of competitive devaluations that contributed to the Great Depression

***laissez-faire* economics** In French, literally "Let them do." A system of economics purporting that there should be minimal government interference.

liberalism advocates for civil liberties for the individual, for limited government and for a free market.

marginal productivity the amount of productivity gained or lost

DOI: 10.1057/9781137062499

marginal utility the amount of pleasure or satisfaction gained or lost from an increase or decrease in the consumption of a commodity

Millennium Ecosystem Assessment A broad study that was overseen by the United Nations (UN). The study addresses ecosystems worldwide.

monetary value The ways in which water is typically used and how it serves financial purposes. Monetary values can be symbolized through units of currency. Once they are symbolized, these values can be exchanged or traded in numeric ways, even in future time. Money is not only a means of exchange, but it also stores value.

opportunity cost the value of the next desirable use (of a commodity) that one must give up if the first option is chosen

organic value Based on the Peoplehood model, organic value emanates from relationships with nature that extend beyond productivity or cash value. Such values often derive meaning through associations with sacred places in the environment.

Pareto Optimal a tenet of new welfare economics which holds that an economic situation is optimal when giving to one individual would not make another worse off

periphery/peripheral countries countries that have a peripheral focus on production

Prior Appropriation Water rights that are tied absolutely to property rights. Sometimes prior appropriation is defined as "first in time, first in right."

riparian rights rights to water are based on the idea that upstream users have first rights

semi-periphery/semi-peripheral countries countries that lie somewhere between the core and periphery of production

supply-side economics A theory of the discipline of economics that supports the idea that supply or output is the basis for growth. Supply-side economics is generally grounded in macroeconomics.

sunk costs expenditures that have already been incurred

sunk loss fallacy the idea that misconceptions of costs can lead to poor decision making

tragedy of the commons an economic problem in which each individual attempts to reap the greatest use from a particular resource. This often causes ecological degradation.

use-value Karl Marx's term that describes the ability of a commodity to "serve the conveniences of human life." *See also* exchange-value.

DOI: 10.1057/9781137062499

usufructuary rights Rights that allow a person to use and enjoy the advantages of another person's property. Privilege of that use is tied to good behavior.

utility the pleasure or satisfaction derived from a good or service

utility discount rate A measure of how much pleasure or satisfaction will be derived from saving a commodity for future use. In other words, a low utility discount rate discounts (disregards) the future slightly, while a high utility discount rate discounts (disregards) the future heavily.

Welfare Economics a branch of economics that addresses social welfare

World Bank Originally known as International Bank for Reconstruction and Development (IBRD), the bank was meant to assist with reconstruction after WWII.

world-systems theory developed by Immanuel Wallerstein to describe the relationship between capitalist production and hegemonic power (*See also* core countries, semi-periphery/semi-peripheral countries, and peripheral countries)

World Trade Organization (WTO) an organization created in 1995 from the General Agreement on Tariffs and Trade (GATT)

DOI: 10.1057/9781137062499

Bibliography

"20 Projects to Showcase 20 Historic Years of Environmental Finance." Millennium Ecosystem Assessment Online Brochure. United Nations Environment Programme (UNEP). Accessed January 10, 2012. http://www.unep.org/dgef/Portals/43/news/facts/GEF%20Folder%20Inserts_12.pdf.

Aaronson, Susan Ariel. "From GATT to WTO: The Evolution of an Obscure Agency to One Perceived as Obstructing Democracy." Economic History Association (February 5, 2012). Accessed February 25, 2013. www.eh.net/encyclopedia/articel/aaronson.gatt.

Abrams, P. "Water in Religion." *The Water Page* (2000). Water Policy International Ltd. Accessed November 30, 2004. http://www.thewaterpage.com/religion.htm.

"African Religion." The Drum. Spring 1995. Accessed December 24, 2004. http://www.ritesofpassage.org/ds95-2.htm.

Ahuvia, Aaron. "If Money Doesn't Make Us Happy, Why Do We Act As If It Does?" *Journal of Economic Psychology*, Vol. 29, Issue 4, August, 2008, pp. 491–507.

"Amrit Ceremony." Religions. BBC, October 27, 2009. Accessed December 20, 2012. http://www.bbc.co.uk/religion/religions/sikhism/ritesrituals/amrit.shtml.

"Amrita." Wikipedia, December 10, 2012. Accessed December 20, 2012. http://en.wikipedia.org/wiki/Amrita.

Anderson, Terry L. and Peter J. Hill, eds. *Water Marketing: The Next Generation*. Lanham, MD: Rowman & Littlefield, 1997.

Anderson, Terry L. and Donald R. Leal. *Free Market Environmentalism.* San Francisco: Pacific Research Institute for Public Policy; Boulder: Westview Press, 1991.

Anderson, Terry L. and Pamela Snyder. *Water Markets: Priming the Invisible Pump.* Washington, D.C.: CATO Institute, 1997.

"Antiquity of Jainism." Jainworld.com. 2011. Accessed December 20, 2012. http://www.jainworld.com/jainbooks/antiquity/jainorel.htm.

Aubuchon, V. *Vaughn's One-Page Summaries.* Vaughn's One-Page Summaries. Accessed January 25, 2005. http://www.vaughns-1-pagers.com/#religion.

"Bahá'í." Adherents.com: National and World Religion Statistics, 2007. Accessed December 21, 2012. http://www.adherents.com/Religions_By_Adherents.html#Baha'i.

"Balinese Hinduism: A Life of Ritual and Devotion." Indonesia-Bali, 2004. Accessed November 30, 2004. http://www.indonesia-bali.com/bali_hindu.htm.

Barnett, Michael and Martha Finnemore. *Rules for the World.* Ithaca, NY: Cornell University Press, 2004.

Beder, Sharon. *Global Spin: The Corporate Assault on Environmentalism.* White River Junction, VT: Chelsea Green Publishing Company, 1998.

Bianchini, John. "'To' bee iina" Water is life." *Navajo-Hopi Observer*, 2004. Accessed November 30, 2004. www.navajohopiobserver.com/news1.htm.

Billi, A., Canitano, G. and A. Quarto. "The Economics of Water Efficiency: A Review of Theories, Measurement Models and Integrated Models," OPTIONS médерranneennes, Series B, n° 57.

Bishop, R. C., Brown M. Welsh, and K. Boyle. "Grand Canyon and Glen Canyon Dam Operations: An Economic Evaluation," in *W-133 Benefits and Costs in Natural Resources Planning, Interim Report #2*. Dept. of Agricultural and Resource Economics, University of Maine: Orono, 1989, http://www.ecosystemvaluation.org/contingent_valuation.htm#case3.

Boas, Taylor C. and Gans-Morse, Jordan. "Neoliberalism: From New Liberal Philosophy to Anti-Liberal Slogan." *Studies in Comparative International Development*. Vol. 44, Issue 2, June 2009: 137–161.

Bowker, J. (ed.). *The Oxford Dictionary of World Religions.* Oxford: Oxford University Press, 1997.

DOI: 10.1057/9781137062499

"Bretton Woods Conference, 1944." U.S. Department of State, Office of the Historian, 2013. Accessed February 24, 2013. http://history.state.gov/milestones/1937–1945/BrettonWoods.

Brauman, Kate A., Daily, Gretchen C., T. Ka'eo Duarte, and Harold A. Mooney. "The Nature and Value of Ecosystem Services: An Overview Highlighting Hydrologic Services." *Annual Review of Environment and Resources*, Vol. 32, Issue 1, 2007: 67–98.

Brown, T.C. "The Concept of Value in Resource Allocation." *Land Economics*, Vol. 60 (1984), 231–246.

"Ceremony of the Blessing of the Water—Churorhnek." Armenian Heritage, 2004. Accessed November 30, 2004. http://www.armenianheritage.com/rewater.htm.

"Chalchiuhtlicue." *Encyclopædia Britannica*. Encyclopædia Britannica Premium Service, (2004). Accessed December 23, 2004, http://www.britannica.com/eb/article?tocId=9022262.

"Chinese Traditional Religion." Adherents.com: National and World Religion Statistics, 2007. Accessed December 19, 2012. http://www.adherents.com/Religions_By_Adherents.html#Chinese.

Cleaver, Frances and Diane Elson. "Women and Water Resources: Continued Marginalisation and New Policies." Sustainable Agriculture Programme of the International Institute for Environment and Development, 1995. Accessed May 15, 2013. http://www.mekonginfo.org/assets/midocs/0003073-environment-women-and-water-resources-continued-marginalisation-and-new-policies.pdf.

Committee on Economic, Social and Cultural Rights, Twenty-ninth Session. "Substantive Issues Arising in the Implementation of the International Covenant on Economic, Social and Cultural Rights" (General Comment No. 15), Geneva, Switzerland, November 11–29, 2002. Accessed September, 2011. http://www.unhchr.ch/tbs/doc.nsf/0/a5458d1d1bbd713fc1256cc400389e94/$FILE/G0340229.pdf.

"Confucianism." ReligionFacts, 2012. Accessed December 19, 2012. http://www.religionfacts.com/a-z-religion-index/confucianism.htm.

"Christianity." Adherents.com: National and World Religion Statistics, 2007. Accessed December 21, 2012. http://www.adherents.com/Religions_By_Adherents.html#Christianity.

Dahl, Arthur Lyon. "The Bahá'í Perspective on Water." *International Environment Forum*, April 18,1999. Accessed January 25, 2005. http://www.bcca.org/ief/ddahl97e.htm.

DOI: 10.1057/9781137062499

Daly, H. and J. Farley. *Ecological Economics: Principles and Application.* Washington, D.C.: Island Press, 2004.

DeFries, Ruth, Pagiola, Stefano, Adamowicz, W. L., Akçakaya, H. Resit, Arcenas, Babu, Suresh, Balk, Deborah, Confalonieri, Ulisses, Cramer, Wolfgang, Fander Falconı´, Steffen Fritz, Rhys Green, Edgar Gutiérrez-Espeleta, Kirk Hamilton, Racine Kane, John Latham, Emily Matthews, Taylor Ricketts, Tian Xiang Yue Augustin, Neville Ash, and Jillian Thönell,. "Analytical Approaches for Assessing Ecosystem Condition and Human Well-being." Millennium Ecosystem Assessment. Washington: Island Press, 2005. Accessed January 10, 2012. http://www.unep.org/maweb/documents/document.271.aspx.pdf

Deurer, R. "Creation Mythology." *Egypt Art* (1997). Accessed November 30, 2004 http://members.aol.com/egyptart/crea.html.

Dore, Mohammed H. I. and Timothy D. Mount, eds. *Global Environmental Economics: Equity and the Limits to Markets.* Malden, MA: Blackwell Publishers Ltd., 1999.

"Draft Environmental Impact Statement for Arizona Snowbowl Facilities Improvement." *Coconino National Forest.* Accessed December 24, 2004. www.fs.fed.us/r3/coconino/publications/snowbowl/.

Ekelund, Robert B. and Robert D. Tollison. *Microeconomics*, 2nd edn. Glenview, IL: Scott, Foresman and Company, 1988.

Ericksen, Polly, Woodley, Ellen, Cundill, Georgina, Reid, Walter V., Vicente, Luís, Raudsepp-Hearne, Ciara, Mogina, Jane, and Per Olsson. "Using Multiple Knowledge Systems: Benefits and Challenges." Millennium Ecosystem Assessment. Washington: Island Press, 2005. Accessed January 10, 2012. http://www.unep.org/maweb/documents/document.343.aspx.pdf.

Espeland, Wendy Nelson. *The Struggle for Water: Politics, Rationality, and Identity in the American Southwest.* Chicago: University of Chicago Press, 1998.

"Estimating the Cost of Disease." *World Health Organization.* World Health Organization, 2004. Accessed June 28, 2004. http://www.who.int/peh/burden/articleEHP052002.pdf.

Femia, Joseph V. *Gramsci's Political Thought: Hegemony, Consciousness, and the Revolutionary Process.* Oxford: Clarendon Press, 1981.

"Flood Myths." *Morgana's Observatory,* 2004. Accessed January 25, 2005. http://www.dreamscape.com/morgana/titania.htm.

DOI: 10.1057/9781137062499

Folke, Carl, Fabricius, Christo, Cundill, Georgina, Schultz, Lisen, Queiroz, Cibele, Gokhale, Yogesh, Marín, Andrés, Camac-Ramirez, Esther, Chandola, Shivani, Tawfic Ahmed, Mohamed, Bibhab Talukdar, Alejandro Argumedo, and Fabricio Carbonell Torres. "Communities, Ecosystems, and Livelihoods." Millennium Ecosystem Assessment. Washington: Island Press, 2005. Accessed January 10, 2012. http://www.unep.org/maweb/documents/document.349.aspx.pdf.

Foster, John Bellamy. *The Vulnerable Planet: A Short Economic History of the Environment*. New York: Monthly Review Press, 1994.

Foster, John Bellamy. *Marx's Ecology: Materialism and Nature*. New York: Monthly Review Press, 2000.

Frederick, Kenneth G. "Preface," pp. vii–ix in Gibbons, Diana C. *The Economic Value of Water*. Washington, D.C.: Resources for the Future, 1986.

Garner, Robert, Ferdinand, Peter, and Stephanie Lawson. *Introduction to Politics*. New York: Oxford University Press, 2009.

Gibbons, D.C. *The Economic Value of Water*. Washington, D.C.: Resources for the Future, 1986.

Goulde, J. "Water in Classical Chinese Religion." *Water: The Mystery, Art and Science of Water.* Sweet Briar College, 2004. Accessed November 30, 2004. http://witcombe.sbc.edu/water/introduction.html.

Gramsci, Antonio. *Selections from the Prison Notebooks of Antonio Gramsci.* New York: International Publishers, 2005.

Grigg, Neil S. *Water Resources Management: Principles, Regulations, and Cases*. San Francisco: McGraw-Hill, 1996.

Guin, M. "Ganga: River and Goddess." *Dolls of India,* 2004. November 30, 2004. http://www.dollsofindia.com/ganga.htm.

Gwartney, James D. "Supply-Side Economics." Library of Economics (2008). Accessed October 17, 2012. http://www.econlib.org/library/Enc/SupplySideEconomics.html.

Hanley, Nick and Clive L. Spash. *Cost-Benefit Analysis and the Environment*. Brookfield, VT: Edward Elgar Publishing Company, 1993.

Hexham, I. *Concise Dictionary of Religion*. Regent College Publishing. University of Calgary, 2000. Accessed January 25, 2005. http://www.acs.ucalgary.ca/~nurelweb/books/concise/.

"Hinduism." Adherents.com: National and World Religion Statistics, 2007. Accessed December 19, 2012. http://www.adherents.com/Religions_By_Adherents.html#Hinduism.

DOI: 10.1057/9781137062499

Holling, C. S., Gunderson, Lance H. and Donald Ludwig. *Panarchy*. Island Press, Washington, D.C., 2002, chapter 1.

Holm, Tom, Pearson, Diane J. and Ben Chavis. "Peoplehood: A Model for the Extension of Sovereignty in American Indian Studies." *Wicazo Sa Review*, Vol. 18, Issue 1, University of Minnesota Press, Spring 2003: 7–24. http://www.jstor.org/stable/1409431.

Holy Bible, Revised Standard Version. Ed. Division of Christian Education of the National Council of the Churches of Christ in North America. Camden, NJ: Thomas Nelson, 1946.

Howard, Charles D.D. and P. Eng. "The Economic Value of Water." *Conference: Mountains as Water Towers* (November 2003). cddhoward.com. Accessed January 13, 2005. http://www.cddhoward.com/docs/Economic%20Value%20of%20Water.pdf.

Hundley Jr., Norris. *The Great Thirst: Californians and Water, 1770s-1990s*. Berkeley: University of California Press, 1992.

Huyler, Jerome. *Locke in America: The Moral Philosophy of the Founding Era*. Lawrence: University of Kansas, 1995.

International Monetary Fund. "Factsheet: The IMF at a Glance" (2013). Accessed February 24, 2013. http://www.imf.org/external/np/exr/facts/glance.htm.

"Islam." Adherents.com: National and World Religion Statistics, 2007. Accessed December 21, 2012. http://www.adherents.com/Religions_By_Adherents.html#Islam.

"Judaism." Adherents.com: National and World Religion Statistics, 2007. Accessed December 21, 2012. http://www.adherents.com/Religions_By_Adherents.html#Judaism.

Ki-Moon, Ban. "Address as Prepared for Delivery to the Davos World Economic Forum," January 24, 2008. Accessed September, 2011. http://www.un.org/apps/news/infocus/sgspeeches/search_full.asp?statID=177.

Kopytoff, Igor. "The Cultural Biography of Things: Commoditization as [P]rocess" in *The Social Life of Things: Commodities in a Cultural Perspective*. Ed. Appadurai, Arjun. Cambridge: Cambridge University Press, 1986.

Laitos, Jan G. and Joseph P. Tomain. *Energy and Natural Resources Law in a Nutshell*. St. Paul: West Publishing Co., 1992.

Lindgren, Anna. "The Value of Water: A Study of the Stampriet Aquifer in Namibia." *Department of Economics, UMEA University*. UMEA

DOI: 10.1057/9781137062499

University (1999). Accessed January 13, 2005. www.econ.umu.se/MFS/annali.pdf.

"Living Beyond Our Means—Natural Assets and Human Well-being: Statement from the Board." Board of the Millennium Ecosystem Assessment. World Resources Institute, March 2005. Accessed January 10, 2012. http://www.wri.org/publication/millennium-ecosystem-assessment-living-beyond-our-means-natural-assets-and-human-we.

Locke, John. *Second Treatise of Government*. Indianapolis: Hackett Publishing Company, Inc., 1980.

Logan, Michael F. *Desert Cities: The Environmental History of Phoenix and Tucson*. Pittsburgh, PA: University of Pittsburgh Press, 2006.

Ludwig, D. "The Era of Management is Over." *Ecosystems*, Vol. 4, 2001: 758–764.

"MA Conceptual Framework." Millennium Ecosystem Assessment. Washington: Island Press, 2005.

Malhotra, K.C., Gokhale, Y., S. Chatterjee, and S. Srivastava. *Cultural and Ecological Dimensions of Sacred Groves in India*. Indian National Science Academy, 2001.

Marx, Karl. *Capital: A Critical Analysis of Capitalist Production*, vol. 1. London: Lawrence & Wishart, Ltd., 2003.

McKinnon, S. "Cities Push Water Conservation: Use Less without Sacrifice." *The Arizona Republic*. AzCentral.com (January 6, 2005). Accessed January 25, 2005. http://www.azcentral.com/specials/special26/articles/0105conserve-main06.html.

McCormick, John. *Reclaiming Paradise: The Global Environmental Movement*. Bloomington, IN: Indiana University Press, 1991.

Millennium Ecosystem Assessment. "MA Conceptual Framework." Washington: Island Press, 2005.

Millennium Ecosystem Assessment (MA). "Ecosystems and Human Well-Being: Synthesis." Washington: Island Press, 2005. Accessed January 10, 2012. http://books.google.com/books?hl=en&lr=&id=_wfumENwOFgC&oi=fnd&pg=PR13&dq=Millennium+Ecosystem+Assessment+(MA).+%E2%80%9CEcosystems+and+Human+Well-Being:+Synthesis.%E2%80%9D&ots=JTgKW3DQGl&sig=PCKR7P4vverATJTGckxOljiEPiw#v=onepage&q=Millennium%20Ecosystem%20Assessment%20(MA).%20%E2%80%9CEcosystems%20and%20Human%20Well-Being%3A%20Synthesis.%E2%80%9D&f=false

DOI: 10.1057/9781137062499

Millennium Ecosystem Assessment (MA). "Ecosystems and Human Well-being: Opportunities and Challenges for Business and Industry." (Washington: Island Press, 2005). Accessed January 10, 2012. http://www.unep.org/maweb/documents/document.353.aspx.pdf.

Muller, A. "A Flower in Full Blossom? Ecological Economics at the Crossroads Between Normal and Post-normal Science." *Ecological Economics*, Vol. 45, 2003:19–27.

Murillo, C. "Steelhead Water Wars." *Santa Barbara Independent,* (Santa Barbara, CA), December 11–18, 2003. Accessed January 31, 2005. http://www.independent.com/cover/cover.html.

Myers, David G. "The Funds, Friends, and Faith of Happy People," *American Psychologist*, Vol. 55, Issue1, January 2000: 56–67.

Myers, David G. and Ed Diener. "Who Is Happy?" *Psychological Science*, Vol. 6, Issue 1, January 1995: 10–19.

Nadeau, Robert L. *The Environmental Endgame: Mainstream Economics, Ecological Disaster, and Human Survival.* New Brunswick, NJ: Rutgers University Press, 2006.

Norgaard, R.B. "The Case for Methodological Pluralism." *Ecological Economics*, Vol. 1, Issue 1, 1989: 37–58.

Online Etymology Dictionary, "Commoditization, 2001–2012." Accessed September 2, 2012. http://www.etymonline.com/index.php?allowed_in_frame=0&search=commodification&searchmode=none.

Paelke, Robert. *Environmentalism and the Future of Progressive Politics.* New Haven, CT: Yale University Press, 1989.

Pagdee, A. Y., Kim, Y. S., and P.J. Daugherty. "What Makes Community Forest Management Successful: A Meta-study from Community Forestry Throughout the World." *Society and Natural Resources*, Vol. 19, Issue 1, 2006: 33–52.

"Pakistan Link." "Issues on Recycled Water and Copyrights." Pakistan Link, 2004. Accessed November 30, 2004. http://www.pakistanlink.com/religion/99/06-04.html.

Pearce, David. "Working Paper: An Intellectual History of Environmental Economics." U.S. Environmental Protection Agency (November 1, 2002). Accessed November 25, 2012. http://yosemite.epa.gov/ee/epa/wpi.nsf/09133da7fb9a95db85256698006641d1/372307b3d147371585256fda004acb6b!OpenDocument.

DOI: 10.1057/9781137062499

Pigou, Arthur C., *The Economics of Welfare*. Library of Economics and Liberty (1932). Accessed November 24, 2012. http://www.econlib.org/library/NPDBooks/Pigou/pgEW1.html.

"Primal-indigenous." Adherents.com: National and World Religion Statistics, 2007. Accessed December 21, 2012. http://www.adherents.com/Religions_By_Adherents.html#primal.

"Rastafari Movement." *Wikipedia*, December 20, 2012. Accessed December 27, 2012. http://webcache.googleusercontent.com/search?q=cache:Cu4QTVfL3g4J:en.wikipedia.org/wiki/Rastafari_movement+rastafarian+symbols+water&cd=7&hl=en&ct=clnk&gl=us.

Robinson, B. A. "Native American Spirituality." *Ontario Consultants on Religious Tolerance*, December 8, 2004. Accessed November 30, 2004. http://www.religioustolerance.org/nataspir.htm.

Rothfeder, J. *Every Drop for Sale: Our Desperate Battle over Water in a World About to Run Out*. New York: Penguin Putnam, Inc., 2001.

Russo, Kira Artemis. Personal Interview, Canada. October 16, 2012.

Ruth, M. "A Quest for the Economics of Sustainability and the Sustainability of Economics." *Ecological Economics*, Vol. 56, 2006: 332–342.

Samuelson, P.A. *Economics: An Introductory Analysis*. New York: McGraw-Hill, 1967.

Sassoon, Anne Showstack, ed. *Approaches to Gramsci*. London: Writers and Readers Publishing Cooperative Society, Ltd., 1982.

Scott, W. Richard. "Approaching Adulthood: The Maturing of Institutional Theory." *Theory and Society*, Vol. 37, Issue 5, October 2008: 427–442.

"Shinto." Adherents.com: National and World Religion Statistics, 2007. Accessed December 19, 2012. http://www.adherents.com/Religions_By_Adherents.html#Shinto.

Shiva, Vandana. *Water Wars: Privatization, Pollution, and Profit*. Cambridge: South End Press, 2002.

Smith, Zachary. *The Environmental Policy Paradox*, 4th edn. Upper Saddle River, NJ: Prentice Hall, 2004.

Subramuniyaswami, Satguru Sivaya. *Dancing with Siva: Hinduism's Contemporary Catechism*. Himalayan Academy Publications, 2003.

Sutaria, Hansa and Vinod Sutaria. "Jain Rituals and Ceremonies." Accessed December 20, 2012. http://www.fas.harvard.edu/~pluralsm/affiliates/jainism/workshop/Sutaria%20Jain%20Rituals.pdf.

DOI: 10.1057/9781137062499

Swatos, Jr., W. *Encyclopedia of Religion and Society*. Hartford Institute for Religion Research, 2005. Accessed January 25, 2005. http://hirr.hartsem.edu/ency/index.html#B.

Tabb, William K. *Economic Governance in the Age of Globalization*. New York: Columbia University Press, 2004.

"Taoism/Daoism." *Yakrider*, 2003. Accessed November 30, 2004. http://www.yakrider.com/Tao/Taoism_Daoism.htm.

Tarlock, A. Dan. "The Future of Prior Appropriation in the New West." *Natural Resources Journal* 41, 2002.

Torras, Mariano and James K. Boyce. "Income, Inequality, and Pollution: A Reassessment of the Environmental Kuznets Curve." *Ecological Economics*, Vol. 25, 1998: 147–160. Accessed March 2013. ftp://131.252.97.79/Transfer/ES_Pubs/ESVal/EnviroKuznetCurve/torras_kuznetandequality_1998_ecolecon_v25_p147.pdf.

"Tsalagi Religion." Sovereign Amonsoquath Tribe of Cherokee, 1994. Accessed November 30, 2004. http://amonsoquath.addr.com/tsalagi_religion.html.

UN Water Decade Programme on Advocacy and Communication (UNW-DPAC). "The Human Right to Water and Sanitation: Milestones," 2011. Accessed September, 2011. http://www.un.org/waterforlifedecade/pdf/human_right_to_water_and_sanitation_milestones.pdf.

United Nations Department of Economic and Social Affairs (UNDESA). "International Decade for Action 'Water for Life' 2005–2015." Accessed May 15, 2013. http://www.un.org/waterforlifedecade/gender.shtml.

Venkatachalam. L. "Environmental Economics and Ecological Economics: Where They Can Converge?" *Ecological Economics*, Vol. 61, 2007: 550–558.

Vörösmarty, Charles J., Lévêque, Christian, Revenga, Carmen, Bos, Robert, Caudill, Chris, Chilton, John, Douglas, Ellen M., Meybeck, Michel, Prager, Daniel, Patricia Balvanera, Sabrina Barker, Manuel Maas, Christer Nilsson, Taikan Oki, and Cathy A. Reidy. "Fresh Water." Millennium Ecosystem Assessment. (Washington: Island Press, 2005). Accessed January 10, 2012. http://www.webpages.uidaho.edu/uiferl/pdf%20reports/MA%20Freshwater%20Ecosystem%20Services.pdf.

Wallerstein, Immanuel. *The Modern World-System: Capitalist Agriculture and the Origins of the European World-Economy in the Sixteenth Century*. New York: Academic Press, 1976.

DOI: 10.1057/9781137062499

Wallerstein, Immanuel. *The Modern World System II: Mercantilism and the Consolidation of the European World-Economy, 1600–1750*. New York: Academic Press, 1980.

"Water Sector Policy Review and Strategy Formulation: A General Framework." *Food and Agriculture Organization of the United Nations*. Food and Agriculture Organization of the United Nations, Rome (1995). Accessed November 30, 2004 http://www.fao.org/docrep/v7890e/V7890E00.htm#Contents.

Whitehead, Alfred North. *Science and the Modern World*. NY: Free Press, 1997.

"Wicca and Witchcraft." ReligionFacts, December 21, 2012. Accessed December 27, 2012. http://www.religionfacts.com/neopaganism/paths/wicca.htm.

Wilkinson, Charles F. *Crossing the Next Meridian: Land, Water, and the Future of the American West*. Washington, D.C.: Island Press, 1992.

Women's Earth Alliance. "Water." Accessed May 15, 2013. http://globalwomenswater.org/. World Bank. "History" (January 31, 2012). Accessed February 24, 2013. http://web.worldbank.org/WBSITE/EXTERNAL/EXTABOUTUS/0,,contentMDK:20653660~menuPK:72312~pagePK:51123644~piPK:329829~theSitePK:29708,00.html.

Worster, Donald. *Rivers of Empire: Water, Aridity and the Growth of the American West*. New York: Pantheon Books, 1985.

Yandle, Bruce, Vijayaraghavan, Maya and Madhusudan Bhattarai. "Income and the Race to the Top." *You Have to Admit It's Getting Better*. Stanford, CA: *Hoover Institution Press*, 2004, pp. 83–108.

"Zoroastrianism." Adherents.com: National and World Religion Statistics, 2007. Accessed December 21, 2012. http://www.adherents.com/Religions_By_Adherents.html#Zoroastrianism.

DOI: 10.1057/9781137062499

Index

DOI: 10.1057/9781137062499

DOI: 10.1057/9781137062499

DOI: 10.1057/9781137062499

DOI: 10.1057/9781137062499

DOI: 10.1057/9781137062499